Biotreatment of Agricultural Wastewater

Editor

Mark E. Huntley, Ph.D.

Associate Research Biologist and Adjunct Lecturer
Scripps Institution of Oceanography
Deputy Director, Water Research Project
University of California, San Diego
La Jolla, California

CRC Press
Taylor & Francis Group
Boca Raton   London   New York

CRC Press is an imprint of the
Taylor & Francis Group, an **informa** business

T0175117

Library of Congress Cataloging-in-Publication Data

Biotreatment of agricultural wastewater/editor, Mark E. Huntley.
    p. cm.
  Bibliography: p.
  Includes index.
  ISBN 0-8493-6378-0
  1. Sewage—Purification—Biological treatment. 2. Agricultural
Wastes. I. Huntley. Mark E., 1950—
TD755.B488   1989
628.1'684—dc19   88-32995

A Library of Congress record exists under LC control number: 88032995

PREFACE

In early 1985 I learned from an article in *Science*[1] that the Kesterson Reservoir, a National Wildlife Refuge approximately 100 miles south of San Francisco which had inadvertently become part of California's vast San Joaquin Valley agricultural drainage system, had become contaminated with concentrations of selenium sufficiently high to result in mass mortality of fish and birds. This event was widely unanticipated,[2-4] but was soon a major story in the world press. In early 1986, a U.S. Department of Interior study identified nine similar sites in six western states where selenium was of concern.[5]

It appeared from all accounts that dissolved selenium was being removed from waters of the reservoir by naturally occurring algae and was from there being passed throughout the food web to wreak havoc among the populations of higher predators. I had long been interested in the mass culture of microalgae, and it seemed to me that this was a case where such technology could be beneficially applied. During the following months, in an effort to discuss this possibility with representatives of relevant state and federal agencies, I met with David Kennedy, Director of the State of California Department of Water Resources, Jesse Hough, Director of Finance of the state, several state legislators, representatives of the U.S. Bureau of Reclamation, and many scientists who were also concerned with the Kesterson problem.

It is not surprising that agricultural wastewater problems in California are not limited to selenium contamination; other metals, nitrates, and salinization in general make significant contributions to the dilemma. The situation is grave; irrigation return water threatens to reduce California's arable land by 20% by the turn of the century,[6] and agricultural regions around the world may face similar problems.

Considering the many advances in biology in the past decade, it certainly seemed that the time was right for an integrated discussion — involving scientists, legislators, administrators, and businessmen — of the role which biotechnology might play in decontaminating agricultural wastewater. With this as the premise, a workshop was organized and held at Scripps Institution of Oceanography August 3 to 5, 1987. Chapters appearing in this book resulted either from manuscripts which authors brought to the meeting or from manuscripts written as a result of working group discussions.

The meeting which resulted in this book would never have taken place if it were not for the substantial contributions of a number of agencies. Primary contributors ($3000 or more) included

Bureau of Reclamation, U.S. Department of Interior
California Sea-Grant College Program, University of California
Alternate Technology Section, Toxic Substances Control Division, Department of Health Services, State of California
Department of Water Resources, State of California
Institute of Marine Resources, University of California
University of California Toxic Substances Research and Training Program, University of California, Davis

Additional contributions were also made by Microbio Resources, Inc., Woodward/Clyde Environmental Consultants (both of San Diego), and by the Office of the Director, Scripps Institution of Oceanography. Without Director Frieman's initial contribution, expressing his strong support for the concept, it would have been difficult to solicit subsequent contributions. I sincerely thank the many people whose commitments of time and effort contributed greatly at various junctures: Thyra Fleming, Mai Lopez, Nicholas Medeiros, Sue Stultz, and Paul Sykes. Finally, many thanks are due to Florence Escritor, who helped at many critical points along the organizational path to assure an orderly and successful meeting, adequate communications between participants, and timely coordination and editing of the manuscripts.

Mark Huntley
Scripps Institution of Oceanography
La Jolla, California
February 1988

REFERENCES

1. **Marshall, E.,** Selenium poisons refuge, California politics, *Science,* 239, 144, 1985.
2. **Marshall, E.,** San Joaquin flooded with water researchers, *Science,* 230, 920, 1985.
3. **Brown, R. L. and Beck, L. A.,** Subsurface agricultural drainage in California's San Joaquin Valley, in *Biotreatment of Agricultural Wastewater,* Huntley, M., Ed., CRC Press, Boca Raton, FL, 1989, chap. 1.
4. **Lee, E. W.,** Current options in treatment of agricultural wastewater, in *Biotreatment of Agricultural Wastewater,* Huntley, M., Ed., CRC Press, Boca Raton, FL, 1989, chap. 3.
5. **Marshall, E.,** High selenium levels confirmed in six states, *Science,* 227, 111, 1985.
6. **Kennedy, D.,** personal communication, November 1985.

The following persons attended the meeting:

Dr. Farooq Azam, Research Microbiologist, Scripps Institution of Oceanography, A-018, University of California, San Diego, La Jolla, CA 92093

Mr. Louis Beck, District Chief, San Joaquin District, Department of Water Resources, 3374 East Shields Avenue, Fresno, CA 93726

Dr. James Blackburn, Assistant Director, Waste Management Research and Education Institute, University of Tennessee, 327 South Stadium Hall, Knoxville, TN 37996-0710

Mr. Charles R. Booth, President, Biospherical Instruments, 4901 Morena Boulevard #1003, San Diego, CA 92117

Dr. Randall Brown, Water Quality Biologist, Department of Water Resources, Central District, 3251 S Street, Sacramento, CA 95816

Dr. Richard Cassin, Research Molecular Biologist, Helicon Foundation, 4622 Santa Fe Street, San Diego, CA 92109

Mr. Tom Collins, Associate Director, Scripps Institution of Oceanography, A-010, La Jolla, CA 92093

Mr. James Costa, State Assemblyman, California State Assembly, State Capitol, Sacramento, CA 95814

Dr. Ronald Crawford, Professor and Head, Department of Bacteriology and Biochemistry, University of Idaho, Moscow, ID 83843

Dr. Joël de la Noüe, Professor, Centre de Recherche en Nutrition, Université Laval, Québec, Québec, Canada G1K 7P4

Dr. Eirik Duerr, Research Biologist, Oceanic Institute, Waimanalo, HI 96795

Mr. Dexter Gaston, President, Microbio Resources Inc., 6150 Lusk Boulevard, San Diego, CA 92121

Dr. Richard Gersberg, Professor of Microbiology, San Diego State University, San Diego, CA

Dr. Mark Huntley, Research Biologist, Scripps Institution of Oceanography, University of California, San Diego, La Jolla, CA 92093

Mr. Franklin ("Pitch") Johnson, President, Asset Management Company, 2275 East Bayshore Drive, Palo Alto, CA 94303

Dr. Roger Korus, Professor of Chemical Engineering, University of Idaho, Moscow, ID 83843

Dr. Edwin Lee, Chief, Technology Development Section, San Joaquin Valley Drainage Program, U.S. Bureau of Reclamation, MP-190, 2800 Cottage Way, Sacramento, CA 95825

Dr. Patrick Mayzaud, Director, INRS Oceanologie, 310 Ave des Ursulines, Rimouski, Québec, Canada G5L 3A1

Dr. Arthur Nonomura, Visiting Scholar, Scripps Institution of Oceanography, La Jolla, CA 92093

Dr. William Oswald, Professor of Sanitary Engineering and Public Health, College of Engineering, University of California, Berkeley, CA 94720

Mr. Steve Pearson, Director, Process-Engineering Group, Woodward/Clyde Consultants, 3467 Kurtz Street, San Diego, CA 92110

Dr. Donald Redalje, Research Biologist, Center for Marine Science, National Space Technology Center, University of Southern Mississippi, NSTL, MS 39529

Dr. Gary Sayler, Professor, Graduate Program in Ecology, University of Tennessee, Knoxville, TN 37916

Dr. Edward Schroeder, Professor of Civil Engineering, University of California, Davis, CA

Dr. Gedaliah Shelef, Professor of Environmental Engineering and Biotechnology, Dean of the Faculty of Civil Engineering, Israel Institute of Technology, Technion, Haifa 32-000, Israel

Dr. Edmund Sperkowski, Transtech Trading Co., 5440 Morehouse Drive, Suite 102A, San Diego, CA 92121

Mr. Paul Stephens, Robertson, Colman and Stephens, 1 Embarcadero Center, Suite 3100, San Francisco, CA 94111

Dr. Aristides Yayanos, Scripps Institution of Oceanography, A-002, University of California, San Diego, La Jolla, CA 92093

THE EDITOR

Mark E. Huntley, Ph.D., is Associate Research Biologist and Adjunct Lecturer, Marine Biology Research Division, Scripps Institution of Oceanography, La Jolla, CA, and Deputy Director, Water Research Project, University of California, San Diego.

Dr. Huntley graduated in 1976 from the University of Victoria, Victoria, British Columbia, with a B.Sc. degree in Biology (first class honors). He obtained his Ph.D. degree in biological oceanography in 1980 from Dalhousie University, Halifax, Nova Scotia. After doing postdoctoral work at the Institute of Marine Resources, Scripps Institution of Oceanography, he was appointed Assistant Research Biologist. In 1984 he moved to the Marine Biology Research Division, where he was appointed Associate Research Biologist in 1987.

Dr. Huntley is a member of the American Society of Limnology and Oceanography, as well as the honorary society Sigma Xi. He has been a recipient of a NATO postdoctoral fellowship (1980 to 1982). He has served in an advisory capacity to the Mayor's Sewer Task Force of the City of San Diego and to the Marine Review Committee of the California Coastal Commission. He was senior scientific advisor to Encyclopedia Brittanica Films for a film entitled "Plankton and the Open Sea" (1985). In 1987 he led a 4-month oceanographic expedition to the Antarctic. He is also president of Aquasearch, Inc., a research and development corporation specializing in the technology of algal mass cultures (1984 to 1988).

Dr. Huntley has been the recipient of grants and contracts from the National Science Foundation, the Office of Naval Research, the Solar Energy Research Institute, and the Sea-Grant College Program. He is the author of more than 30 papers and has presented more than 50 lectures at national and international meetings, institutes, and universities throughout the world. His current research interests include the physiological ecology of zooplankton, the effects of plankton on optical properties of seawater, and the application of algal mass culture technology to wastewater treatment and the production of chemicals and pharmaceuticals.

CONTRIBUTORS

Louis A. Beck, B.S.
Chief, San Joaquin District
California Department of Water
 Resources
Fresno, California

James W. Blackburn, Ph.D.
Research Associate Professor
Department of Chemical Engineering
Center for Environmental Biotechnology
Energy, Environment, and Resources
 Center
University of Tennessee
Knoxville, Tennessee

Randall L. Brown, Ph.D.
Environmental Program Manager
California Department of Water
 Resources
Sacramento, California

Richard C. Cassin, Ph.D.
Research Member of the Foundation
Department of Ocean Sciences
Helicon Foundation
San Diego, California

Paris H. Chen, Ph.D.
Senior Engineer and Technical Consultant
R.C.M., Inc.
Methane Recovery Systems
Berkeley, California

Ronald L. Crawford, Ph.D.
Professor and Head
Department of Bacteriology and
 Biochemistry
University of Idaho
Moscow, Idaho

Joël de la Noüe, D.Sc.
Professor
Department of Biology
University of Laval
Ste.-Foy, Quebec, Canada

Eirik O. Duerr, Ph.D.
Research Scientist and Program Manager
Aquaculture Research Service Program
The Oceanic Institute
Waimanalo, Hawaii

Matthew B. Gerhardt, M.S.
Research Assistant
Sanitary Engineering and Environmental
 Health Research Laboratory
University of California
Berkeley, California

F. Bailey Green, M.S.
Energy and Resources Group
University of California
Berkeley, California

Mark E. Huntley, Ph.D.
Associate Research Biologist
Marine Biology Research Division
Scripps Institution of Oceanography
La Jolla, California

Franklin P. Johnson, Jr.
Lecturer in Management
Graduate School of Business
Stanford University
Stanford, California

Edwin W. Lee, E.D./D.P.A.
Supervisory Environmental Engineer
San Joaquin Valley Drainage Program
U. S. Bureau of Reclamation
Sacramento, California

Patrick Mayzaud, D.Sc.
Director
Department of Oceanography
I.N.R.S.
Rimouski, Quebec, Canada

Robert D. Newman, B.A.
Junior Associate
Department of Civil Engineering
SHEERL
University of California
Berkeley, California

Arthur M. Nonomura, Ph.D.
Chief Technological Advisor
United Nations Development Program
United Nations
New York, New York

Yakup Nurdogan, Ph.D.
Chemical Engineer
Bechtel Environmental, Inc.
San Francisco, California

Kirk T. O'Reilly, Ph.D.
Department of Bacteriology and
  Biochemistry
University of Idaho
Moscow, Idaho

William J. Oswald, Ph.D.
Professor
Departments of Civil Engineering and
  Public Health
University of California
Berkeley, California

Steven Pearson, B.S.
Vice President
Woodward-Clyde Consultants
San Diego, California

Donald G. Redalje, Ph.D.
Assistant Professor
Center for Marine Science
University of Southern Mississippi
John C. Stennis Space Center
SSC, Mississippi

Gary S. Sayler, Ph.D.
Professor and Director
Center for Environmental Biotechnology
University of Tennessee
Knoxville, Tennessee

Gedaliah Shelef, Ph.D.
Professor and Dean
Department of Civil Engineering
Israel Institute of Technology
Technion
Haifa, Israel

Leslie Shown, M.S.
Junior Associate
College of Engineering
University of California
Berkeley, California

Christine S. Tam
Engineering Aid
College of Engineering
University of California
Berkeley, California

David F. Von Hippel, Ph.D.
Research Associate
Energy Systems Research Group
Boston, Massachusetts

TABLE OF CONTENTS

Chapter 1

SUBSURFACE AGRICULTURAL DRAINAGE IN CALIFORNIA'S SAN JOAQUIN VALLEY

Randall L. Brown and Louis A. Beck

TABLE OF CONTENTS

I. INTRODUCTION

The objective of our chapter is to set the stage for subsequent presentations and discussions in this book by describing the history of the San Joaquin Valley drainage problem and the environmental issues surrounding it. The description focuses on drainage from California's San Joaquin Valley; however, the material on environmental concerns and salt balance is relevant to other areas of the world where similar problems exist.

The San Joaquin Valley (Figure 1) has long been one of the world's richest agricultural areas. In 1979, the value of crops produced in the valley was estimated at $7 billion.[1]

California's Mediterranean climate results in winter rains and hot, dry summers. This climate pattern and the long growing season cause valley farmers to rely on irrigation. Irrigation water is derived from groundwater pumping, transfer from storage reservoirs in the Sierra Nevada mountains to the east, and state and federal water projects that bring in water from the Sacramento-San Joaquin Delta to the north.

At present, about 2 million ha of the San Joaquin Valley are irrigated with about 7 km³ of water annually. In much of the valley, irrigation causes few direct environmental problems, although there are concerns associated with the various diversions used to provide the needed water. On about 200,000 ha, however, irrigation has resulted in problems caused by poor drainage. These lands, located mainly on the valley's west side, are underlain with almost impermeable clay layers that severely limit the downward penetration of water. Depending on the hydraulic gradient, portions of the applied water can accumulate in the root zone beneath the irrigated fields or can flow downslope and cause similar problems on the valley floor. Evapotranspiration from plant surfaces and dissolution of native soil minerals cause the salt content of the percolating water to increase to levels where it is not easily recycled as irrigation water.

The most common solution to the drainage problem on the west side has been to install artificial drains at field depths of 2 to 3 m. Figure 2 illustrates how these permeable drains collect the perched water and lower the water table to below the root zone. About 50,000 ha of land with high water tables in the San Joaquin Valley have been drained in this fashion.

This physical solution has transferred the salty water from beneath the farmers' fields to other areas. In California's Imperial Valley, which borders on Mexico to the south, about 200,000 ha of farmland have been drained in a similar manner. The drainage goes to a confined natural salt sink, the Salton Sea, which for years has sustained California's most productive inland fishery. Because of the sea's high salt content, the fishery consists of marine species imported from the Gulf of California plus such salt-tolerant species as tilapia.

As long as the amount of land drained in the San Joaquin Valley remained relatively modest, much of the drainage in the northern valley flowed to the San Joaquin River and out to sea by way of the Sacramento-San Joaquin Delta. (Water from the southern valley could not take this pathway because of a natural slight rise in the valley floor which usually hydraulically isolates the Tulare Lake basin from the San Joaquin River.) However, as drainage volumes built up, there were valid concerns that beneficial uses of the San Joaquin River would be degraded by discharge of too much salt and possibly other harmful materials to the river. These concerns led to several attempts to find other means of resolving drainage-related issues. The remainder of this chapter describes the issues themselves and summarizes the various formulations developed to respond to the issues.

II. DRAINAGE WATER QUALITY AND QUANTITY

Irrigation water applied to the fields is generally of high quality, with concentrations of total dissolved solids less than 500 mg/l. Evapotranspirative processes result in pure water leaving the plant surfaces, with the residual salts being left in the soil profile. Farmers apply

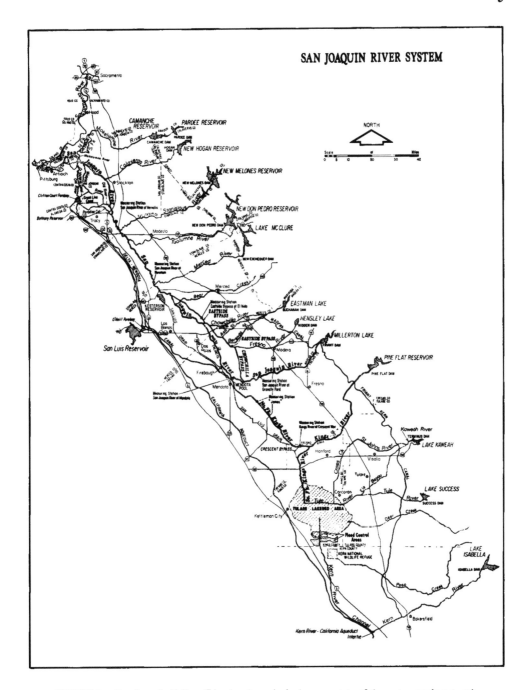

FIGURE 1. San Joaquin Valley, CA, showing principal components of the water supply network.

extra water (the leaching fraction) to move the deposited salts away from the root zone. As
the water percolates through the soil, it also dissolves native minerals. The result of both
actions is an appreciable increase in salt content and a change in the ionic composition and
trace mineral content of the leachate.

The exact changes in quality depend on irrigation practices, soil characteristics, and
length of time the land has been drained. For example, drainage water collected from
individual systems can have total dissolved solids concentrations ranging from less than
2,000 mg/l to more than 100,000 mg/l. Generally, the drainage is less salty in the north,

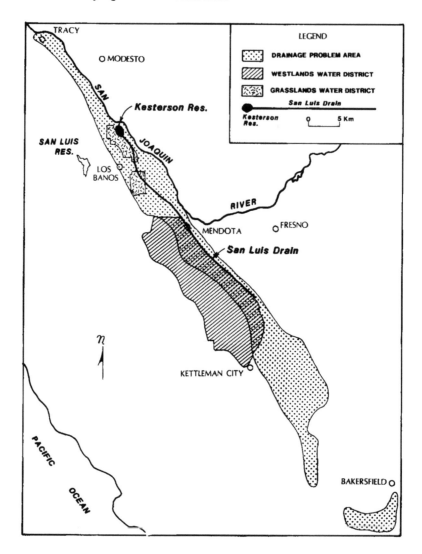

FIGURE 2.   Drainage problem area and Kesterson Reservoir, located in the western central portion of the San Joaquin Valley, CA.

moderate in the south (Tulare Lake basin), and highest in the central portion of the valley's west side. Concentrations of nitrate and boron follow the same general pattern.

There have been several attempts to predict the quality of drainage that would result if the effluents from different areas were combined in a single collector system. Four such independent estimates of mineral quality are shown in Table 1. These estimates are for drains that would have been in operation for some time (up to 50 years). Although purely hypothetical, the estimates themselves are in reasonable agreement considering the number of assumptions that were required for the calculations. A few points can be made regarding these predictions.

1.   The water is moderately brackish, with a salinity about one fifth that of sea water.
2.   The drainage is a sodium sulfate type.
3.   Nitrate-nitrogen concentrations are quite high when compared to those found in the applied water. (Nitrate is essentially the only inorganic nitrogen ion present.)
4.   Phosphorus occurs in relatively low concentrations as compared to nitrogen.
5.   High concentrations of boron, a phytotoxin, are present, which can interfere with reusing drainage in irrigation.

TABLE 1
Comparison of Various Estimates of Steady-State Chemical Composition of Subsurface Drainage from the San Joaquin Valley after Several Years of Operating a Combined Drainage System

Constituent	IDP[a]	DWR[b]	Price[c]	USBR[d]
Calcium and magnesium	6200	380	500	664
Sodium	970	1900	1500	1370
Potassium	5.3	20	20	6
Bicarbonate	350	220	200	350
Sulfate	1750	3500	3000	3279
Chloride	660	1000	1200	826
Boron	10.2	11	10	9
Nitrate-nitrogen	21	20	20	25
Total phosphorus	0.31		0.15	0.10
Total dissolved solids	6200	6800	7000	6600

Note: Concentrations in mg/l.

[a] Interagency Drainage Program (IDP)[2] — projected values for the year 2025.
[b] California Department of Water Resources (DWR)[3] — projected Master Drain after 50 years of operation.
[c] Price[4] — projected for San Luis Drain, 1970.
[d] U.S. Bureau of Reclamation (USBR)[5] — San Luis Drain, 2020 estimated quality.

It must be emphasized that the values in Table 1 are for idealized composite samples — the water from any particular area might be higher or lower than shown. For the central area, for example, average total dissolved concentrations approach 10,000 mg/l, with the concentration of the various ions increasing accordingly.

In addition to common chemicals, drainage water contains appreciable amounts of trace elements. Table 2 lists the ranges of trace elements reported by the State of California Department of Water Resources (DWR) and the U.S. Bureau of Reclamation (USBR) for the federal San Luis Drain. This drain collected water from the central portion of the valley's west side, so the data are only representative of this area. The table shows a fairly wide range of trace elements in the drainage. Levels of selenium, strontium, chromium, iron, and nickel are particularly elevated. If the drainage samples analyzed had been from the Tulare Lake basin, arsenic concentrations would be higher (up to 1 mg/l in some drainage systems) and selenium levels would be lower. From a treatment perspective, the important point is that there may be more than one constituent of environmental concern.

Farmers in the west side of the San Joaquin Valley apply over 150 organic chemicals to soils and crops for pest control.[6] In general, these compounds are not found at high concentrations in subsurface drainage, although they are often found in tailwater (surface runoff) from irrigated fields. Even the persistent chlorinated hydrocarbons (such as DDT, dieldrin, and endrin) are decreasing in subsurface drainage (Table 3). The relatively low pesticide concentrations in subsurface drainage are probably due to a combination of degradation (enhanced by long travel time through the soil) and adsorption onto soil particles.

Finally, subsurface drainage has little turbidity and few readily available organic compounds. The water is slightly alkaline as it enters the collector drains and is relatively well buffered by bicarbonate alkalinity.

Exact quantities of drainage from the San Joaquin Valley are difficult to estimate. The Interagency Drainage Program[2] estimated that drainage yield varied from 0.5 to 1.2 acre-

TABLE 2
Range in Dissolved Trace
Element Concentrations (μg/l) in
Waters from the San Luis Drain,
1983—1984

Element	Range (μg/l)
Selenium	230—350
Strontium	6400—7200
Iron	150—500
Cadmium	1—20
Chromium	20—36
Copper	10—20
Lead	1—6
Manganese	10—20
Mercury	0.1—0.2
Nickel	20—60
Silver	1
Zinc	10—20

Note: Analysis by the California Department
of Water Resources and the U.S. Bureau of Reclamation.

TABLE 3
Concentration of Various Chlorinated
Hydrocarbons Detected in Subsurface Agricultural
Drainage from California's San Joaquin Valley
during the Period 1970—1982

Compound	Maximum concentration (μg/l) in time interval		
	1978—1982	1975—1977	1970—1974
Aldrin	0.01	0.025	0.025
DDD/DDE/DDT	ND[a]	0.17	80
Dieldrin	ND[a]	0.01	25
Endrin	0.01	0.02	3
Toxaphene	ND[a]	0.54	335
PCBs	ND[a]	0.19	0.19

[a]   Not detectable.

ft/acre (average = 0.8), depending on location and age of the drainage systems. If all problem lands were drained, upwards of 600,000 acre-ft of drainage would be produced annually. Drainage facilities (canals and pipelines) designed to transport combined valley drainage have varied from 300 to 900 ft^3/s capacity, depending on the size of the area being served.

III. DRAINAGE WATER MANAGEMENT

An appreciation of present-day treatment concerns can be obtained from brief summaries of the various drainage management scenarios advanced over the past 40 years and the needs these scenarios were designed to alleviate.

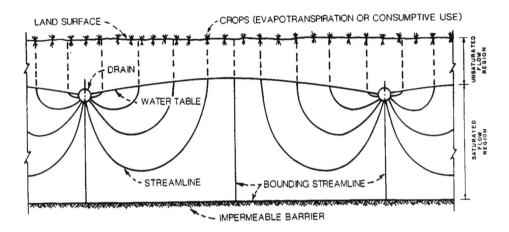

FIGURE 3. Vertical section through a typical agricultural drainage field.

A. SAN LUIS DRAIN — ORIGINAL PLANNING

In the early 1950s, when engineers began planning for delivery of Central Valley Project water to the valley's west side, they knew artificial drainage would be essential to prevent rising water tables downslope of the irrigators and to remove imported salts from the valley. The San Luis interceptor drain (Figure 3) was authorized as a project feature. The drain was to collect water from the federal service area and discharge it to the delta near Antioch. Project planning did not advance to the stage where a detailed analysis of environmental concerns was prepared.

B. SAN JOAQUIN MASTER DRAIN

In the late 1950s, as the State of California began planning to bring irrigation water to the San Joaquin Valley, its engineers and agricultural scientists also recognized the need for drainage facilities. The primary purpose of the proposed facilities was to provide a means for removing as much salt from the valley as was being delivered in irrigation water imported to the valley. A detailed analysis of means for disposing of this water was conducted. Options evaluated included desalting, direct discharge to the ocean, evaporation ponds, and discharge to the delta near Antioch. In the preliminary version of the final report,[3] the recommended project was to combine the state and federal drains (to be called the Master Drain), with discharge to the delta. A report on expected problems[7] in receiving waters singled out nitrogen and pesticides as constituents of particular concern.

In the late 1960s, the Federal Water Pollution Control Administration (now the Environmental Protection Agency) released a report[8] describing studies to assess the impact of a delta discharge on receiving water quality. In essence, federal investigators concluded that the high nitrogen levels in the effluent could cause adverse levels of algal growth (eutrophication) in the estuary. They recommended that nitrogen removal be part of any planned discharge. Pesticides were also considered in the analysis, but levels in the drainage were found to be similar to those already present in the delta. Since no particular problems had been identified as being caused by ambient delta concentrations, pesticides were not included as an area of particular concern.

As a result of concerns about potential eutrophication problems in the delta, the DWR, the USBR, and the Federal Water Pollution Control Administration initiated a cooperative 3-year treatment study. Most of the studies were conducted at a DWR field laboratory on the west side of the valley and included investigations of desalting, bacterial denitrification, and algal stripping. Results, as summarized by Brown,[9] demonstrated that

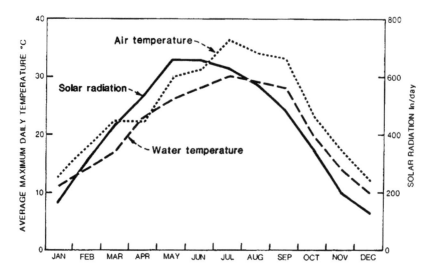

FIGURE 4.   Annual cycle of monthly mean solar insolation and air and water temperatures at Firebaugh, central San Joaquin Valley, CA.

1.   Desalting by electrodialysis or reverse osmosis could effectively remove salts, but did little to remove boron or nitrate.
2.   Bacterial denitrification, using methanol as a carbon source, could be used to lower the nitrate concentration from 20 to about 2 mg/l. Upflow rock reactors seemed to be the most promising container; however, possible hydraulic problems associated with plugging by actively growing bacteria were never completely resolved. Required residence time was in the range of 2 to 4 h, and methanol dosages were about 60 mg/l when nitrate-nitrogen concentrations were 20 mg/l.[10]
3.   High densities of algae could be grown in outdoor cultures by using relatively long residence times (5 to 16 d), shallow depths (20 to 40 cm), mixing, and additions of small amounts of phosphorus and iron. Ferric chloride was used as a coagulant to harvest the algal mass. The effluent from the ponds contained about 3 to 5 mg/l total nitrogen, mostly as dissolved and particulate organic nitrogen. The San Joaquin Valley's Mediterranean climate and drainage discharge patterns combined to make the use of outdoor cultures particularly attractive, since the seasonal periods of highest light intensity coincided with periods of maximum drainage volume. Figure 4 illustrates a typical annual cycle of air temperature, solar radiation, and water temperature at a California Irrigation Management Information System site in the central west side of the valley. Figure 5 shows that drainage flows are also generally higher during the months when sunlight and temperature are maximum.

Preliminary cost estimates (1970 dollars) developed for the two biological processes were about $24/1000 m³ for the bacterial process and about $36/1000 m³ for the algal stripping process. The algal process would require about 5000 ha of ponds to treat the estimated flow of the San Joaquin Master Drain.[11] Larger scale desalting studies were needed to work out pretreatment requirements for these processes before cost estimates could be made for the processes.

In 1969, the State of California determined that it could not finance its portion of the Master Drain and withdrew from the project.

C. SAN LUIS DRAIN — INITIAL CONSTRUCTION
To fulfill its continuing obligation to provide drainage facilities, in 1968 the USBR

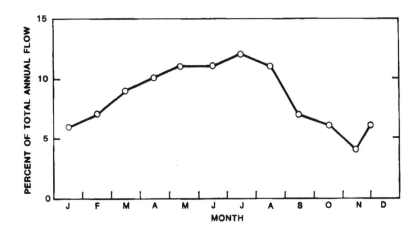

FIGURE 5. Percentage of total annual drainage flow, on a monthly basis, in the federal service area, central San Joaquin Valley, CA.

began construction of the middle 130 km of the 320-km San Luis Drain to the delta. This section of the drain was completed in 1974 and terminated in an interim holding facility, the Kesterson Reservoir. Although over 2000 ha were purchased for the reservoir, only about 500 ha were developed into a series of about 40-ha cells. These cells were surrounded by levees and had an average depth of 1.2 to 1.5 m. They were designed to provide interim storage until completion of the drain to the delta, with some of the storage coming from evaporation losses. The ponds were built more like marshes than true evaporation ponds in that water depth was quite variable.

Environmental concerns related to a delta discharge and to financing caused the USBR to postpone its plans to complete the San Luis Drain. Structural integrity of the concrete drain itself was maintained by keeping it full of good quality water. Similar quality water also was released into Kesterson Reservoir, which promoted the growth of emergent vegetation and helped convert the reservoir to a brackish water marsh. The U.S. Fish and Wildlife Service began to manage the reservoir as part of its Kesterson National Wildlife Refuge to provide additional waterfowl habitats in the San Joaquin Valley.

In the early 1970s, the USBR, with technical assistance from the DWR, continued biological studies of removing nitrogen from subsurface drainage.[12] The main area of investigation was the so-called "symbiotic" process, where a combination of algal and bacterial pathways was used to remove nitrogen. The plants (either algae or higher aquatic plants) incorporated some of the dissolved nitrogen into their cells. Far more importantly, the plants also provided an organic carbon source for bacterial denitrification. Although it was difficult to demonstrate conclusively, bacterial denitrification apparently occurred in microlayers at the pond bottom or in decomposing plant mass and was the major pathway by which inflowing nitrogen was removed from drainage water. The symbiotic process appeared technically feasible in that about 90% nitrogen removal was obtained. It was also economically attractive because neither mixing nor an outside organic carbon source was required. As a sidelight, it was found that periphytic diatoms attached to higher aquatic plants (or an inert substrate such as dead tumbleweeds) effected considerable silicon removal. Silicon can cause scaling when drainage is used in power-plant cooling or in desalting plants. Ponds of alkali bullrush were subsequently tested for their potential in pretreating drainage going to a demonstration-scale reverse osmosis desalting facility and were shown to be effective at removing silicon.

D. INTERAGENCY DRAINAGE PROGRAM (THREE-AGENCY)

In 1975, the USBR, DWR, and the State Water Resources Control Board formed the

Interagency Drainage Program (IDP) in another attempt to find an environmentally acceptable and cost-effective means to resolve the valley's drainage problem.

Much of the work was developed around the principles and standards of the Water Resources Council.[13] The goal was to develop separate environmental and economic plans and then merge the two into a recommended plan. The final plan[2] was a valley drain with discharge to the western delta at Chipps Island. Unlike the San Joaquin Master Drain, the IDP drain included several thousand hectares of waterfowl habitat (marshes) and regulating reservoirs. The California Department of Fish and Game and the U.S. Fish and Wildlife Service participated in the program and endorsed the recommended plan.

With respect to problem chemical constituents, mathematical modeling indicated that a delta discharge would cause only local increases in dissolved nitrogen and algal growth. Thus, the plan did not include nitrogen removal requirements, but did indicate that nitrogen removal might be needed if the actual discharge caused unacceptable increases in algal growth in the estuary. Arsenic from Kern County was listed as a concern; however, source control appeared to be the most likely option for avoiding receiving water problems.

In his comments on the plan's environmental impact report, Dr. Ray Krone, a consultant for the Contra Costa Water District, pointed out that selenium might also be a trace constituent of concern. Dr. Krone, from the University of California, Davis, was concerned that selenium levels might be elevated in subsurface drainage. These concerns were based on high selenium concentrations in some groundwater near Davis and the possibility that similar selenium-rich formations existed further down the valley.

There were no formal treatment studies underway during the late 1970s, although the DWR and the faculty and staff from the aquaculture program at the University of California, Davis, investigated the potential use of agricultural drainage as a culture medium for growing economically important invertebrates and fish. As reported by Monaco et al.,[14] most of the species tested did well in the drainage. Exceptions were the Asiatic clam and the Pacific crayfish, but the data were not adequate to determine the causes of the problems to these animals. During this period, Oswald[15] made an extensive analysis of projected changes in water quality as the water passed through the marshes, regulating reservoirs, and canals of the IDP's recommended plan. Oswald's report provides much useful information on environmental conditions in the San Joaquin Valley as they relate to the potential for outdoor treatment systems.

The IDP plan, like the Master Drain, could not be financed by the valley agricultural community, and it was not constructed. As with the Master Drain and the San Luis Drain, there was considerable adverse reaction from agencies, organizations, and individuals around the bay-delta regarding an estuary discharge. In addition, Tulare Lake basin farmers did not actively support the plan, choosing to rely instead on construction of in-valley evaporation ponds.

E. SAN LUIS DRAIN — REPORT OF WASTE DISCHARGE

As before, the USBR still had the legal obligation to alleviate and prevent drainage problems in its water service area. In the late 1970s, tile drainage systems from about the first 20,000 ha provided with collector systems were connected to the San Luis Drain, and by 1981 the entire flow in the drain was from subsurface drainage. The drainage water flowed into Kesterson Reservoir. Thus, for the first time the marsh plants and wildlife were exposed to actual drainage water, with all its potential problem constituents.

In 1980, the USBR began active planning to complete the San Luis Drain to the western delta near Chipps Island. Preparation began on an environmental impact document and a report of waste discharge (as required by the National Pollution Discharge Elimination System).

Both documents had gone through several drafts when the U.S. Fish and Wildlife Service began to express concerns that drainage water was causing waterfowl deaths and deformities at Kesterson Reservoir. Preliminary indications were that selenium, found at concentrations of 300 to 400 μg/l in San Luis drainage water, was the element responsible for water fowl problems.[16] These findings prompted the State Water Resources Control Board to issue Order WQ 85-1, calling for plugging the collector lines conveying drainage to the San Luis Drain (and Kesterson) and the cleanup of Kesterson Reservoir. Planning for a delta discharge was also halted pending resolution of new environmental issues resulting from the Kesterson findings.

F. INTERAGENCY DRAINAGE PROGRAM (FIVE-AGENCY)

In 1984, three federal agencies (U.S. Bureau of Reclamation, U.S. Fish and Wildlife Service, and U.S. Geological Survey) and two state agencies (California Departments of Water Resources and Fish and Game) formed a second Interagency Drainage Program. The general objective of providing an environmentally sound and economically feasible solution to the valley's drainage problems remained unchanged. However, the emotional and technical issues related to selenium had dramatically changed the range of options available to the planners and regulatory agencies. Selenium is found in widely varying concentrations in valley drainage and appears to be leaching from natural shales in the alluvial fans (and interfans) on the valley's west side. Possible bioaccumulation and biomagnification of selenium through the food web complicates the task of setting water quality standards for this element.

The five-agency drainage program includes research by the U.S. Geological Survey on the chemical composition of drainage from various areas in the valley and research by the U.S. Fish and Wildlife Service on effects of selenium and other trace elements on fish and wildlife. Planning focuses on local solutions to drainage problems, including evaluation of possible land retirements in the major problem areas, treatment to remove problem constituents, and more efficient use of drainage water to reduce effluent quantity. Discharge to the ocean, either directly or via the San Joaquin River or the delta, is not politically possible at this time.

IV. SUMMARY AND DISCUSSION

The drainage problem has been around for some time, as has the perceived need to remove something from the water to make discharge of the water to a salt sink (such as the ocean) environmentally acceptable. This "something" has varied from total salts to nitrogen to pesticides to various trace elements, including selenium. As part of its Order WQ 85-1, the State Water Resources Control Board[17] established interim standards for the San Joaquin River for total salts, boron, selenium, and molybdenum. The interim selenium objective of 5 μg/l will likely drop to 2 μg/l after a few years.

Similar standards or objectives for the delta or evaporation ponds have not been adopted. Evaporation ponds are, however, subject to provisions of the California Administrative Code, which defines the levels of particular elements that cause the ponds to be classified as toxic waste pits. In the case of selenium, the ponds become toxic waste pits when concentrations exceed 1 mg/l. Ponds that exceed these values (and there are already a few in the valley) must be closed or double-lined to prevent leakage from ponds. The California Department of Fish and Game is also considering imposing waterfowl hazing programs on ponds containing selenium to minimize the birds' exposure to potential toxic contaminants.

In the Grasslands Water District (Figure 2), duck club managers have long used subsurface drainage as an important part of their water supply. The blended drainage water, which was moderately high in selenium (in the 50 to 100 μg/l range), flowed through the marshes and out to the San Joaquin River. Marsh vegetation and bacteria acted as a biological treatment system and removed much of the selenium as the water passed through the system.

As a result of Order WQ 85-1 and concerns associated with elevated selenium levels in waterfowl, the duck clubs no longer take drainage water. The water now goes directly to the San Joaquin River, and treatment may be required to meet the new water quality objectives established for the river.

Publicity over the Kesterson findings and concern over the possibility of similar problems with wildlife refuges receiving return flows from agriculture and other sources has resulted in the so-called "westwide study". In this study, the U.S. Geological Survey and U.S. Fish and Wildlife Service are conducting reconnaissance-level studies to determine if selenium or other trace elements are causing problems in other refuges throughout the West. In California, the Salton Sea and the Tulare Lake basin are included in the study program, and preliminary results have shown elevated selenium levels in biota from these two areas.

Various studies have demonstrated that it is technically feasible to remove salts, nitrogen, and boron from drainage water. Studies underway now also indicate that selenium can be removed by physical, chemical, or biological processes. The challenge is to provide processes that result in an environmentally safe effluent (i.e., remove selenium as well as other toxic substances), that do not result in residues that also cause problems, and that farmers can afford. We are optimistic that continued scientific research and technology development will help meet this challenge.

REFERENCES

1. **Security Pacific Bank,** A Statistical Profile of the San Joaquin Valley, Security Pacific Bank, Sacramento, CA, 1980.
2. **Interagency Drainage Program,** Agricultural Drainage and Salt Management in the San Joaquin Valley, Final Report Including Recommended Plan and First Stage Environmental Impact Report, Interagency Drainage Program, Fresno, CA, 1979.
3. **California Department of Water Resources,** San Joaquin Master Drain, Preliminary Ed., Bull. No. 127, DWR, Fresno, CA, 1965.
4. **Price, E. P.,** Genesis and scope of interagency cooperative studies of control of nitrates in subsurface agricultural wastewaters, in Collected Papers Regarding Nitrates in Agricultural Wastewaters, Water Pollut. Control Res. Ser. 13030 ELY 12/69, U.S. Environmental Protection Agency, Cincinnati, 1969.
5. **U.S. Bureau of Reclamation,** Draft Report of Waste Discharge — San Luis Drain, Mid-Pacific Region, U.S. Bureau of Reclamation, Sacramento, CA, 1983.
6. **Marine Bioassay Laboratories,** Study Plan Recommendations: Aquatic Toxicity Testing and Comprehensive Monitoring for the San Luis Drain, Central Valley Project, California, prepared for the Mid-Pacific Region, U.S. Bureau of Reclamation, Marine Bioassay Laboratories, Sacramento, 1983.
7. **Oswald, W. J., Crosby, D. D., and Golueke, C. G.,** Removal of Pesticides and Algal Growth Potential from San Joaquin Valley Drainage Waters — A Feasibility Study, submitted to the California Department of Water Resources, Sanitary Engineering Reserach Laboratory, University of California, Berkeley, 1964.
8. **Federal Water Pollution Control Administration,** San Joaquin Master Drain — Effects on Water Quality of San Francisco Bay and Delta, Southwest Region, FWPCA, San Francisco, 1967.
9. **Brown, R. L.,** The occurrence and removal of nitrogen in subsurface agricultural drainage from the San Joaquin Valley, California, *Water Res.*, 9, 529, 1975.
10. **McCarty, P. L.,** Feasibility of the denitrification process for removal of nitrate nitrogen from agricultural drainage waters, in Field Evaluation of Anaerobic Denitrification in Simulated Deep Ponds, Bull. 174-3, California Department of Water Resources, Sacramento, May 1969.
11. **Brown, R. L.,** Removal of Nitrate by an Algal System, Water Pollution Control Res. Ser. 13030 ELY 4/71, U.S. Environmental Protection Agency, Cincinnati, 1971.
12. **Cardon, D., Cedarquist, N., and Rumboltz, M.,** Removal of Nitrate by a Symbiotic Process, Water Pollut. Control Res. Ser. 13030 ELY 11/74-15, U.S. Environmental Protection Agency, Cincinnati, 1975.
13. Principles and Standards, Water Resources Council, Washington, D.C., 1973.
14. **Monaco, G., Brown, R. L., and Gull, G. A. E.,** Exploring the Aquacultural Potential of Subsurface Agricultural Drainage Water, Aquaculture Program, University of California, Davis, 1981.

15. **Oswald, W. J.,** Projected changes in quality of San Joaquin Valley subsurface drainage waters in a proposed marsh and canal system, Final Report to the State Water Resources Control Board, W. J. Oswald & Associates, Berkeley, CA, 1978.

16. **Ohlendorf, H. M., Hothem, R. L., Bynck, C. M., Aldrich, T. W., and Moore, J. F.,** Relationships between selenium concentrations and avian reproduction, *Trans. North Am. Wildl. Nat. Resour. Conf.,* 51, 330, 1986.

17. **State Water Resources Control Board,** Regulation of Agricultural Drainage at the San Joaquin River, Technical Committee Report, Order No. W.Q. 85-1, SWRCB, Sacramento, 1987.

Chapter 2

DECIDING ON A TREATMENT ALTERNATIVE

Steven Pearson

TABLE OF CONTENTS

I. INTRODUCTION

Although irrigation is probably one of the most important practices developed by man, through the centuries it has brought with it some serious problems. The importance of irrigation, however, cannot be overemphasized. It was practiced by the Indians when Columbus arrived and the remnants of these old systems can still be observed.[1] Today, millions of acres of land are irrigated, a practice which has contributed to the high agricultural productivity of the U.S. Unfortunately, although much of the water applied to agricultural lands is used by the crops, some of the water finds its way back into surface and underground water resources. This water, known as agricultural return flow, often introduces into the resource a variety of pollutants that were part of the agricultural operation or were a result of the agricultural erosion of the soil. Typical agricultural wastewater pollutants include dissolved salts of sodium, potassium, calcium, magnesium, chloride, sulfate, nitrate, and phosphates; pesticides, which include herbicides, insecticides, and miticides; trace metals such as selenium, chromium, arsenic, and manganese; and soil particles, which contribute sediment, silt, and turbidity. Agricultural return flow contaminated with these and other pollutants tends to pollute lakes, streams, marshes, underground aquifers, and other natural bodies of water. The damage to water resources and wildlife habitats caused by this pollution is a serious concern. The quantities of agricultural wastewater currently being generated in the U.S. have been reported to be in the billions of gallons each day.[2] The treatment or prevention of agricultural wastewater has been a focus of attention for the last several decades.

Because of the large quantities of water which require treatment and the comparatively low capital and operating costs, biological treatment, in its many forms, has attracted attention as a possible solution to many parts of this important treatment problem. It is not technically feasible for biological treatment to solve all agricultural wastewater problems; economic constraints require that it be applied in those situations where it is the most feasible alternative. Chemical and physical treatments or source control should be considered when biological treatment does not appear to meet treatment needs. There are many types of established biological systems, each with different capabilities, and new and innovative technologies are developed and tested each year.

The biological system must be selected properly and an understanding achieved of what it can and cannot do. Realistic expectations of the selected biological treatment system and the successful integration of chemical and physical treatments as components of the system, where appropriate, are often keys to the overall success of the treatment program. The proper treatment option should be determined through a well-thought-out process that thoroughly investigates the specific situation and all available alternatives.

II. TREATMENT OPTIONS — AN OVERVIEW

The general treatment alternatives available for the treatment of agricultural wastewater can be divided into two major treatment categories: (1) physical/chemical treatment systems and (2) biological treatment systems (Figure 1). The process leading to a decision to pursue biological treatment must be part of an overall evaluation of available alternatives that have potential for solving the specific agricultural wastewater treatment problem. The terms "biological treatment" and "physical/chemical treatment" embody such a broad and diverse series of treatment technologies that it is impossible to make a treatment decision based on this general characterization. Some overall comparisons can be made, however, if these broad categories are subdivided into major treatment groups.

A. BIOLOGICAL TREATMENT SYSTEMS

Biological treatment refers to the utilization of living organisms for the purpose of reducing the concentration of contaminants in a waste stream. Although not always the case,

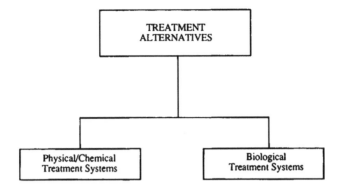

FIGURE 1. General treatment alternatives for agricultural wastewater.

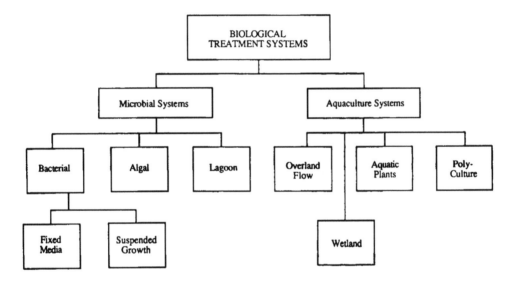

FIGURE 2. Biological treatment system.

many biological treatments take place through the utilization of natural metabolic processes. Organic and inorganic contaminants are used by biological systems to form new cellular material or to produce energy which is required by the organisms for their life systems. Even simple biological systems are complex mixtures of thousands of biochemical reactions being conducted by a variety of biological organisms. This complexity often produces an amazing ability by the biological system to adapt to the treatment of a wide variety of pollutants. While many groupings are possible, biological treatment systems can be divided into two main categories: (1) microbial systems and (2) aquaculture systems (Figure 2).

1. Microbial Systems

Of the microbial systems, bacterial systems, although they often incorporate many other forms of microbial organisms, are probably the best known.[3-7]

The established systems can be divided into two groups: (1) suspended-growth systems and (2) fixed-film systems. The best known suspended-growth bacterial system is the activated sludge process which has been a standard in the treatment of municipal wastewaters for many years. These systems have an established record for the efficient removal of suspended and dissolved organic materials, nutrients, and some trace metals.

The basic system consists of a large basin into which the contaminated wastewater is introduced, along with either air or oxygen, utilizing diffusion or mechanical aeration devices. The microorganisms are present in the aeration basin as suspended material. After the microorganisms remove the contaminants from the wastewater, they are separated from the water by gravity settling. After settling, a portion of the settled biomass is returned to the aeration tank and the remainder is removed for treatment, reuse as a soil amendment, or disposal.

Fixed-film biological systems differ from suspended-growth systems in that microorganisms attach themselves to a medium which provides inert support. Biological towers, trickling filters, and rotating biological contactors are the most common types of fixed-film systems. In these systems, microbes form a slime layer over the inert media and metabolize contaminants as they flow through the system. Aeration is generally provided by moving air at a countercurrent to the water flow or by diffusion from the atmosphere. Fixed-film systems remove from wastewater materials similar in type to those removed by suspended-growth systems.[8]

Lagoon systems are microbial systems that generally rely on the symbiotic relationship between algae and bacteria. In shallow aerobic lagoons, very high treatment efficiencies for organic wastes and nutrients can be achieved if aeration and light penetration allow the maximization of the photosynthetic and bacterial oxidation processes.

Algal systems have received a great deal of attention during the last several years for the removal of nutrients from wastewater.[9,10] In addition, specific algae have been found to metabolize or complex organic and inorganic compounds which were previously thought to be refractory to biological systems.[11-17]

2. Aquaculture Systems

Like microbial systems, aquaculture systems have many forms. Aquaculture generally refers, however, to the utilization of higher forms of plant and animal life for the main components of the biological system. These systems are as varied as nature itself, but can generally be divided into four categories: (1) overland flow, (2) wetland processes, (3) aquatic plant processes, and (4) polyculture.

Overland flow and other land treatment biological systems have been found to be very effective for the removal of nutrients from wastewaters.[18,19] Under controlled conditions and with aerobic conditions in the soil, plant nutrients and other residuals can be removed or degraded by the microorganisms in the soil horizon or by chemical precipitation, ion exchange, biological transformation, and biological uptake through the root systems of the vegetative cover.

Wetland processes include natural or constructed bogs, marshes, and swamps. Water may or may not flow out of the wetland process. A substantial amount of the treatment taking place in wetland processes is due to the bacteria and other microorganisms found on the plant stalks and decaying biological materials. The humic deposits found in these processes are also responsible for the adsorption of contaminants from the wastewater.

Aquatic plant processes are those that utilize high forms of aquatic plant life for a portion of the treatment process. Probably the most frequently studied aquatic plant for wastewater treatment is the water hyacinth,[20-22] although systems utilizing duckweed, water lettuce, water primrose, and others have also been studied.[23,24] While the aquatic plants are responsible for the removal of some nutrients from the wastewater, the most important role of the plant is to provide a substrate for the growth of bacteria and other microorganisms, which probably have the larger role in the treatment process. In addition, some aquatic plants have been reported to remove certain toxic metals from wastewater.[25,26]

Polyculture refers to the utilization of a variety of organisms working in a symbiotic relationship to provide the necessary treatment. These systems are characterized by aquatic plant systems that also incorporate crayfish, fish, insect larvae, and bottom-dwelling worms

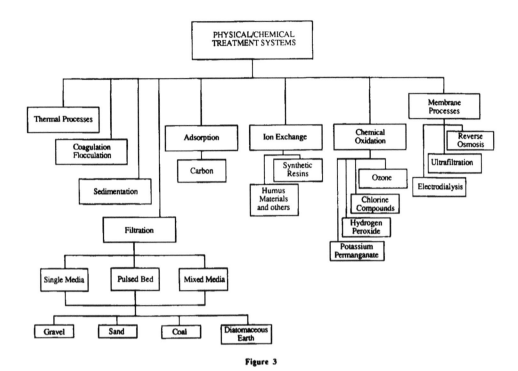

FIGURE 3. Physical/chemical treatment system.

and other aquatic organisms to effect a complete food chain treatment process. One of the reported advantages of these systems is the generation of less sludge since the bacterial biomass is converted to higher forms of life.

Biological systems traditionally have been utilized to treat organic wastes that can be readily metabolized by bacteria or algae and to remove nutrients from wastewater. Trace metal removal has largely been attributed to precipitation processes or adsorption onto coagulated biological particles. Recent studies have shown, however, that some biological systems may remove heavy metals from wastewater in other ways.[25,26] Very few biological systems have shown an ability to reduce salinity or to remove dissolved salts from wastewater.

B. PHYSICAL/CHEMICAL TREATMENT SYSTEMS

Physical/chemical treatment systems utilize physical or chemical processes to remove contaminants from wastewater. They can be grouped into the following eight categories (Figure 3):

- Coagulation and flocculation
- Sedimentation
- Filtration
- Adsorption
- Ion exchange
- Membrane processes
- Chemical oxidation
- Thermal processes

Coagulation and flocculation refer to two modes of colloidal particle destabilization. Some biological treatment systems (i.e., activated sludge) often depend on the coagulation of the biomass for effective settling and removal. The chemical coagulation process, however, utilizes chemical agents such as high-molecular-weight polymers to precipitate and coagulate pollutants into particles which will settle and be removed. This process has been effectively used for the removal of heavy metals from industrial wastewaters.[27,28] Metals, by the addition of a strong base, sulfides, or other anions, are precipitated as their hydroxide or other insoluble salt, coagulated, and removed from the system as metal-bearing sludge.

Sedimentation refers to the settling of particulate or coagulated suspended solids from the wastewater. Sedimentation is often used in conjunction with coagulation and flocculation, although it is also independently effective for suspended particles which are unstable in suspension because of their size and density. Sedimentation is an effective treatment for agricultural wastewaters containing soil particles which have not formed irreversible colloidal suspensions.

Filtration processes involve a wide variety of gravity and pressure filters, including mixed media and pulsed bed filters, which utilize sand, gravel, diatomaceous earth, cloth, and coal as the filtering media. The basic principle lies in the removal of precipitated or coagulated contaminants by physically straining them from the water. The filtration process mechanisms are complex, but the basic principle is simple. Filtration processes have generally been employed as a polishing step following coagulation and sedimentation rather than directly following coagulation.

Adsorption, particularly carbon adsorption, has become a widely used treatment process for the removal of organic compounds from wastewater.[29,30] The carbon adsorption process occurs when the wastewater contaminant molecules or atoms nearly uniformly interpenetrate the carbon atoms. The factors affecting carbon adsorption are surface area (carbon mesh size and type), contaminant type and concentration, and pH. While high removal efficiencies can be achieved for a wide range of organic contaminants, the cost compared to other treatment systems is high due to the cost of high-quality activated carbon.

Ion exchange is a process in which ions, held by electrostatic forces to charged functional groups on the surface of a solid, are exchanged for contaminant ions of similar charge in the wastewater. It is a sorption process where ions make a phase change from solution to attachment to a solid. Ion exchange has been used extensively for the removal of calcium and magnesium ions from wastewater, but has also been used for the removal of chromium, uranium, and other metals from wastewater.[31,32] Synthetic exchange resins are the most common exchange solids, but humus materials, wood, coal, metal oxides, algae, and bacteria are known to support ion-exchange phenomena. Modern synthetic ion-exchange resins can be regenerated by washing the resin with a concentrated solution of the originally held ions.

Membrane processes utilize semipermeable membranes to separate pollutants from the wastewater. The best known membrane processes are reverse osmosis, ultrafiltration, and electrodialysis. Membrane processes, particularly reverse osmosis and electrodialysis, are effective for the removal of dissolved salts and other pollutants from wastewater. They can easily reduce the salt concentration by a factor of 10. These processes, however, require significant pretreatment and are known to be high energy consumers, although significant developments within the last 5 years have reduced the power consumption by nearly 50%. They also produce a concentrated brine waste stream containing the concentrated contaminants, which must receive additional treatment and/or be discarded.

Chemical oxidation is a process in which the oxidation state of the contaminant is raised to a higher level. The purpose behind the oxidation process is to convert the contaminant to a more environmentally acceptable compound. In the case of toxic organic hydrocarbons, for example, the goal is to oxidize them to carbon dioxide and water. Some inorganic compounds, such as cyanides and sulfides, can be effectively oxidized to more desirable compounds. The chemical oxidation agents most commonly employed are ozone, chlorine and chlorine compounds, potassium permanganate, hydrogen peroxide, and oxygen.[34-36] Care must be taken that the chemical by-products are indeed more desirable than the original contaminant. Chemical oxidation systems often have high operating costs which are directly associated with the costs of the chemicals they consume.

Thermal processes include distillation and evaporation, which are designed to remove dissolved solids from the wastewater by driving off the water in the vapor phase. Both processes are effective for the removal of dissolved salts. Distillation is a high energy consumer and has been replaced by reverse osmosis in many applications. Evaporation, including solar evaporation, is generally a slower process requiring less energy, but more area, than distillation, and it is often utilized when the recovery of evaporated water is not important.

The overview of treatment options presented above portrays the myriad of options available to solve a particular wastewater treatment problem. Biological treatment is a viable treatment alternative for many agricultural wastewater contaminants. Biological treatment in combination with a physical/chemical process may present advantages and, in some cases, a pure physical/chemical system is the most feasible alternative available. Arriving at the selected treatment alternative, whether it be a biological or physical/chemical treatment, involves the same process of investigation and evaluation.

III. MAKING THE TREATMENT DECISION

It is important that all potential treatment alternatives be evaluated according to a systematic method which incorporates the uniqueness of the particular agricultural wastewater and the environment in which it exists. While there are many approaches to the evaluation and selection of a treatment alternative, the following system has been used successfully for this type of decision making. It is comprised of modifications to the predefined importance scale models for decision making developed by Linstone, Turoff, Rau, and others.[37,38] The Treatment Evaluation System (TES), as depicted on Figure 4, consists of seven basic steps:

1.   Data collection and analysis
2.   Identify treatment alternatives
3.   Establish screening criteria
4.   Screen treatment alternatives
5.   Establish evaluation criteria
6.   Evaluate feasible treatment alternatives
7.   Rank feasible treatment alternatives

Each of these successive steps incorporates a set of tasks which focuses the treatment on the specific agricultural wastewater, condition, and environment. The goal of the TES is to select a treatment alternative based upon a careful evaluation which has utilized a set of criteria specifically developed for the particular wastewater being considered for treatment.

The TES utilizes a subjective numerical process for evaluating feasible alternatives. A greater degree of confidence can be achieved in the results of this process if two or three people with different perspectives agree on the selection of criteria, assignment of importance factors, and numerical values. Discussion and agreement is part of a successful evaluation.

A. DATA COLLECTION AND ANALYSIS

Data collection and analysis is the process of evaluating available data on the particular agricultural wastewater, determining data gaps, and developing a field sampling program to obtain missing data considered important to the evaluation process (Figure 5). A fairly large general data base is available for the contaminants found in various types of agricultural wastewater.[39,40] For specific cases, however, data may be missing or the reliability may be in question. Characterization of the agricultural wastewater by careful evaluation of new and existing data is an important step which can lead to the success or failure of a treatment technology; ignoring data gaps at this point can come back to haunt the designer of an

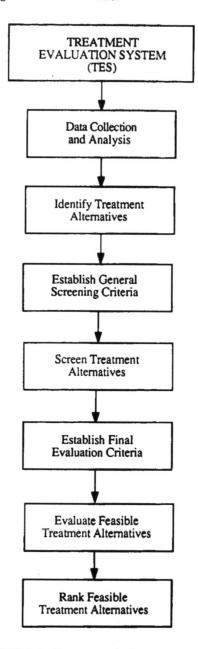

FIGURE 4.   Treatment evaluation system (TES).

agricultural wastewater system. If existing data are judged to be insufficient, the design and execution of a field sampling and analytical program are required before further treatment-alternative evaluation can proceed.

As an example of how the TES can work, treatment alternatives will be examined for a relatively simple case involving an agricultural wastewater with the following character-istics:

Biochemical oxygen demand	60 mg/l
Total suspended solids	350 mg/l
Total volatile suspended solids	60 mg/l
Nitrate nitrogen	35 mg/l
Total phosphorus	0.5 mg/l
pH	6.5

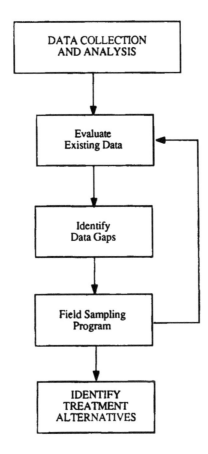

FIGURE 5.   Data collection and analysis.

The agricultural wastewater is one of several discharges flowing into a nearby lake and is suspected of contributing to the overall eutrophication of the lake. Both community groups and regulatory agencies have expressed concern about pollution in the lake and want the discharges, including this one, cleaned up. For our example, we will assume that the above characterization is the result of several years of data collected by county and state agencies and is considered to be reliable. No further data gathering, therefore, is determined to be warranted at this time.

B. IDENTIFY TREATMENT ALTERNATIVES

Identification of treatment alternatives produces a listing of available options for the treatment of the specific agricultural wastewater. The identification process includes a waste treatment characterization, a general review of contaminant treatability, and a listing of direct disposal options (Figure 6). The waste treatment characterization establishes the pollutants of concern, their maximum and average concentrations, and the acceptable residual wastewater concentrations.

The general review of contaminant treatability provides a preliminary review of the appropriateness of the proposed treatment technology for the treatment of the pollutant or pollutants of concern. An assessment of the comparative treatment performance is not made at this stage as long as it can be established that the proposed treatment technology exhibits some promise for the treatability of the target pollutants. In addition, the listed treatment technologies need not exhibit treatability for all of the listed pollutants since some combination of treatment technologies may be selected. The quantity of the wastewater to be

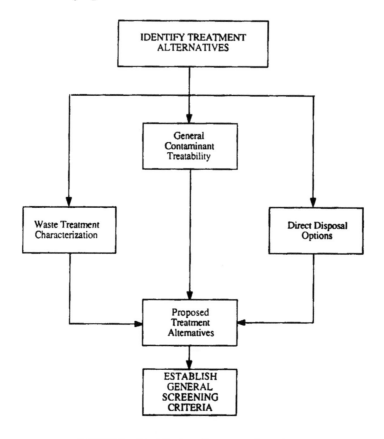

FIGURE 6.   Identification of treatment alternatives.

treated (projected flow) is also considered. Direct disposal options are listed so that they may be considered as an alternative to treatment of the wastewater. The identification of treatment alternatives should be performed during the creative, conceptualization stage of the decision-making process, in which new and innovative ideas are allowed to be expressed. It is important at this stage to guard against the elimination of creative concepts due to preconceived ideas, past experiences, and professional training; treatment alternatives that exhibit unacceptably high risks, uncertainties, or cost factors are eliminated later during the evaluation process.

Using our agricultural wastewater problem, we understand that the regulatory agencies have established the following goals for controlling pollution of the lake:

Biochemical oxygen demand	30 mg/l
Total suspended solids	30 mg/l
Nitrate nitrogen	5 mg/l
Total phosphorus	0.6 mg/l

The average wastewater flow was found to be 50,000 gal/d, with variations of between 20,000 and 100,000 gal/d throughout the year. A review of the treatment goals reveals the need to reduce nitrate nitrogen and total suspended solids. After review of the wastewater data and a good deal of discussion, our three-person evaluation team identifies the following list of proposed treatment alternatives:

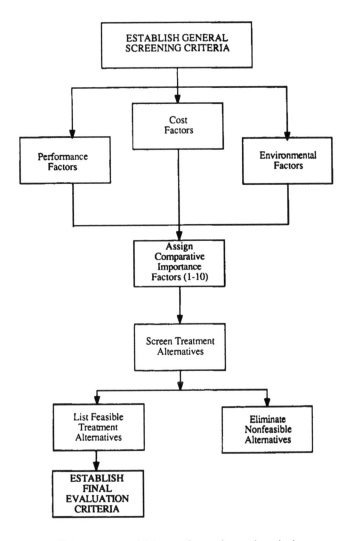

FIGURE 7. Establishment of general screening criteria.

1. No treatment (continued discharge to the lake)
2. Clarification followed by a traditional rotating biological contactor
3. Clarification followed by a traditional suspended-growth biological reactor
4. Clarification by an innovative algae harvesting system
5. Chemical clarification followed by ion exchange
6. Chemical clarification and filtration followed by reverse osmosis

C. ESTABLISH GENERAL SCREENING CRITERIA

The general screening criteria are developed following the process of identifying treatment alternatives. The components of the general screening criteria can be divided into three broad categories: performance, cost, and environmental factors (Figure 7). The purpose for the establishment of screening criteria is the development of a list of factors considered to be vital to the success of the treatment project. The comparative importance of the factors is identified, and a subjective numerical value between 1 and 10 is assigned. Since these are general screening criteria, it is not critically important to have hard data on each treatment alternative, although the better the data or information on the proposed treatment technology, the better the screening.

TABLE 1
Sample General Screening Criteria
Importance Factors

General screening criteria	Importance factor
Capital cost	8
Operating cost	5
Community acceptability	9
Reliability	7
Regulatory acceptability	6
Overall treatment performance	8

Note: Importance factors are on a scale of 1 to 10, with 10
being the highest rating.

Performance factors might include the ability of the treatment technology to achieve the desired discharge concentration and to handle the anticipated influent concentration without adverse effects, the influence of flow variability on performance, compatibility with other treatment technologies, and the availability of established design criteria. Cost factors might include the capital costs associated with the construction of the treatment alternative, operation and maintenance costs, and monitoring costs. Environmental factors might include regulatory controls, effects on human health, residual disposal problems, community acceptability, and native flora and fauna impacts. Any factor considered at this stage should be deemed critical to the project's success; less critical factors will be considered later in the evaluation process.

The general screening criteria established for our agricultural wastewater treatment case example by our committee of three is shown in Table 1.

D. SCREEN TREATMENT ALTERNATIVES

Utilizing the screening criteria established in the first step, each treatment alternative listed is rated for each screening criterion by assigning a numerical value between 1 and 10, with 10 being the best rating. After each of the alternatives has been rated, the rating is multiplied by the comparative importance factor (estimated during the previous step) to obtain the factored criterion score. A treatment alternative score is obtained by totalling the factored criteria scores for each alternative (highest score) ranked number one. At this stage, several alternatives are generally still being considered, but treatment alternatives with very low rankings can be eliminated as nonfeasible. Typically, between three and five alternatives are still being considered as viable at this point.

For our sample agricultural wastewater problem, after establishing the general screening criteria, our committee of three screened each proposed alternative (Table 2). The results of their screening, after multiplying by the general criteria importance factors and totalling, are found in Table 3. Based upon the results of the general screening, the committee eliminated alternatives 1 (no treatment) and 6 (reverse osmosis) from further consideration.

E. ESTABLISH FINAL EVALUATION CRITERIA

The next steps involve a more critical evaluation of the remaining feasible alternatives. Similar to the process utilized to establish the general screening criteria, final evaluation criteria are established which will be utilized to critique the feasibility of the remaining alternatives. The final evaluation criteria can be divided into the following six categories (Figure 8):

- Human health impacts
- Feasibility and performance

TABLE 2
Sample General Screening Criteria Rankings

Treatment alternative	General screening criteria					
	1	2	3	4	5	6
No treatment	10	10	1	1	1	1
Rotating contactor	7	7	8	8	7	7
Suspended-growth reactor	6	5	6	7	8	7
Algae harvesting	8	7	8	8	7	8
Ion exchange	4	5	6	8	5	7
Reverse osmosis	2	3	6	5	5	6

Note: Rankings are on a scale of 1 to 10, with 10 being the highest rating.

TABLE 3
Sample General Screening Criteria
Results

Treatment alternative	Score
Algae harvesting	156
Rotating contactor	148
Suspended-growth reactor	129
Ion exchange	118
Reverse osmosis	93
No treatment	72

Note: The higher the score, the more preferred the treatment.

- Cost feasibility
- Environmental impacts
- Regulatory acceptability
- Timeliness (need for pilot testing)

All criteria considered to be important to the success of the project are included as final evaluation criteria, and a comparative importance factor of between 1 and 10 is assigned to each evaluation criterion. The final evaluation criteria may come from regulators, public interest groups, environmental groups, and other sources in addition to those that are established based on technical performance. Decidedly better results are usually obtained if several individuals, representing a spectrum of perspectives, participate in the establishment of the final evaluation criteria. For our sample problem, final screening criteria were established and importance factors assigned (Table 4).

F. EVALUATE FEASIBLE TREATMENT ALTERNATIVES

The evaluation of feasible treatment alternatives is a critical evaluation that should be based on actual operation data, literature citations, regulator decisions, or previous working experience. The remaining alternatives are rated for each established evaluation criterion by assigning a value of between 1 and 10, with 10 being the best score.

Often, new and innovative ideas do not have a sufficient data base, or more traditional treatments have not been used on the particular wastewater being evaluated. The result of this absence of data can either be low scoring of the proposed treatment alternative or the design and execution of a pilot testing program that seeks to obtain the missing data, as

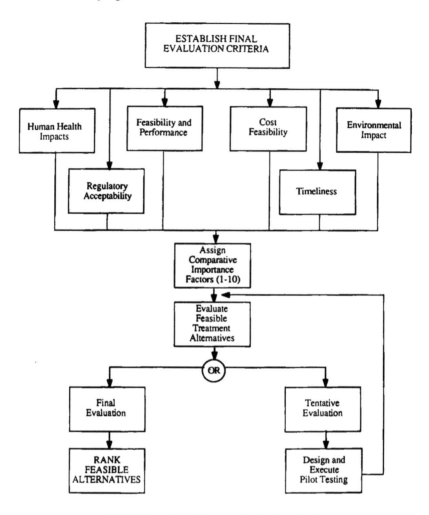

FIGURE 8.   Establishment of evaluation criteria.

TABLE 4
Sample Final Screening Criteria Importance Factors

Final evaluation criteria	Importance factor
Discharge meets groundwater standards	5
Regulatory permitting problems	8
Waste by-product production	5
Waste by-product disposal options	9
Handle variable influent concentrations	6
Chemical use	3
Treatment reliability	7
Treatment effectiveness	9
Capital cost	9
Operating cost	5
Time schedule	7
Odor problems	6
Visual appearance	5

Note: Importance factors are on a scale of 1 to 10, with 10 being the
highest rating.

TABLE 5
Sample Final Evaluation Criteria Rankings

Evaluation criteria	Treatment alternative			
	1	2	3	4
Groundwater standards	3	3	3	3
Permitting problems	6	7	8	5
By-product production	7	6	5	5
By-product disposal	7	7	6	4
Variable influent concentration	8	7	8	8
Chemical use	8	8	8	4
Treatment reliability	8	7	8	7
Treatment effectiveness	9	7	7	7
Capital cost	9	8	6	5
Operating cost	8	8	6	4
Time schedule	4	9	7	8
Odor problems	8	7	6	9
Visual appearance	7	7	6	9

Note: Rankings are on a scale of 1 to 10, with 10 being the highest rating.

TABLE 6
Sample Treatment Alternatives Final Score

Treatment alternative	Final score
Algae harvesting	602
Rotating biological contactor	594
Suspended growth biological reactor	548
Ion exchange	508

Note: The higher the score, the more preferred the treatment.

indicated in Figure 7. The final evaluation may therefore not be completed until the pilot testing is completed. In order to determine if pilot testing is the most feasible option, a tentative ranking is prepared utilizing the evaluator's experience, training, and base knowledge about the various alternatives. If the pilot testing option is determined to be the most feasible alternative, final evaluation of the alternatives is postponed until the pilot testing data have been obtained.

Completion of the evaluation process is accomplished by multiplying the evaluation criterion score for each alternative by the importance factor assigned to the final evaluation criteria. Overall treatment alternative scores are obtained by adding the evaluation criteria scores.

For our sample treatment problem, the evaluation found in Table 5 was provided by the review committee even though good performance data could not be found for the innovative algae harvesting system that they had proposed. Tentative evaluation data were therefore provided for this system.

After the alternatives were scored, each criterion score was multiplied by the importance factor established in the previous step. The results of the scoring are shown in Table 6.

G. RANK FEASIBLE TREATMENT ALTERNATIVES

The feasible treatment alternatives are ultimately ranked in order of feasibility. While this methodology can be utilized for the systematic evaluation of treatment alternatives, the numerical values obtained should be recognized as subjective aids to a decision-making process, of little or no value in the prediction of actual performance. Alternatives with similar numerical values probably should be considered to have similar feasibilities, and one should not be preferred over another. This process can assist with the design of treatment programs by evaluating what factors are critical for treatment success. If budgets allow pilot testing, new technologies can be developed to treat agricultural wastewaters which compare favorably, in a systematic way, to established technologies.

The results of our sample treatment problem indicated that the proposed innovative algae harvesting system was worth pilot testing. The tentative evaluation ranked it best, but this was a subjective evaluation based upon incomplete data. The more traditional rotating biological contactor and suspended-growth biological contactor systems were rated roughly equivalent to the algae harvesting system. Based upon the final evaluation, the ion-exchange concept was dropped from further consideration. After the pilot testing of the algae harvesting system, the final evaluation could be completed and a single alternative selected.

IV. SUMMARY

Many options exist for the treatment of agricultural wastewaters. There exist a number of proven technologies for both organic and inorganic agricultural wastes and several new emerging technologies which hold promise for the future. The best type of treatment is usually dependent upon many factors, all of which may not be readily apparent without careful evaluation. Whatever alternative is selected, however, the treatment costs will be significant. For these reasons, it is wise to undertake a rigorous evaluation of the treatment alternatives available before making a final decision. The TES provides a relatively simple model for such a decision-making process. While the TES model provides an analytical approach to evaluating existing technologies, it can also be utilized to compare them with more innovative approaches that may still require pilot testing. This allows new technologies to survive an engineering assessment without being automatically eliminated as unproven.

REFERENCES

1. **Houk, I. E.,** Irrigation engineering, in *Agriculture and Hydrologic Phases,* Vol. 1, John Wiley & Sons, New York, 1951.
2. **U.S. Department of the Interior,** Characteristics and Pollution Problems of Irrigation Return Flow, U.S. Department of the Interior, Washington, D.C., 1969.
3. **Stainer, R. Y., Ingramham, J. L., and Adelberg, E. A.,** *The Microbial World,* 4th ed., Prentice-Hall, Englewood Cliffs, NJ, 1976.
4. **Young, J. C. and McCarty, P. L.,** The anaerobic filter for waste treatment, *J. Water Pollut. Control Fed.,* Vol. 41, 1969.
5. **Higgins, T. J. and Burns, R. G.,** *The Chemistry and Microbiology of Pollution,* Academic Press, New York, 1975.
6. **Weston, R. F. and Eckenfelder, W. W.,** Application of biological treatment to industrial wastes. I. Kinetics and equilibria of oxidative treatment, *Sewage Ind. Wastes,* 27, 802, 1955.
7. **Metcalf and Eddy, Inc.,** *Wastewater Engineering: Collection, Treatment, Disposal and Reuse,* 2nd ed., Tchobanoglous, G., Ed., McGraw-Hill, New York, 1979.
8. **U.S. Environmental Protection Agency,** Process Design Manual for Nitrogen Control, U.S. EPA, Washington, D.C., 1975.
9. **Stowell, R.,** A Study of the Screening of Algae from Stabilization Ponds, Masters thesis, University of California, Berkeley, 1976.
10. **Albertson, O.,** Nutrient Control, Water Pollution Control Federation, Washington, D.C., 1983.
11. **Gale, N. L. and Wixon, B. G.,** Removal of heavy metals from industrial effluents by algae: development in industrial microbiology, *Soc. Ind. Microbiol.,* 1979.

12. **Krauss, H. J. and Porter, J. W.,** The absorption of inorganic ions by *Chlorella pyrenoidosa, Plant Physiol.*, 29, 234, 1954.
13. **Goldman, J. C. and Ryther, J. H.,** Waste reclamation in an integrated food chain system, in *Biological Control of Water Pollution*, University of Pennsylvania Press, Philadelphia, 1976, 197.
14. **McGriff, E. and McKinney, R.,** The removal of nutrients and organics by activated algae, *Water Res.*, 4(10), 1115, 1972.
15. **Mcknight, D. M., Pereira, W. E., Rostad, C. E., and Stiles, E. A.,** Effect of retorted-oil shale leachate on blue-green algae *(Anabaena flos-aquae), Bull. Environ. Toxicol.*, 30, 6, 1983.
16. **Cerniglia, C. E., Gibson, D. T., and Van Baalen, C.,** Oxidation of naphthalene by cyanobacteria and microalgae, *J. Gen. Microbiol.*, 116, 495, 1980.
17. **Cerniglia, C. E., Van Baalen, C., and Gibson, D. T.,** Oxidation of biphenyl by the cyanobacterium, *Oscillatoria* sp., strain JMC, *Arch. Microbiol.*, 125, 203, 1980.
18. **U.S. Environmental Protection Agency,** Process Design Manual for Land Treatment of Municipal Wastewater, 626/1-77-088, U.S. EPA, Washington, D.C., October 1977.
19. **Doneen, L. D., Ed.,** *Proc. Symp. Agricultural Wastewaters*, Rep. No. 10, Water Resources Center, University of California, Davis, 1966.
20. **Reed, S. C. and Bastian, R. K.,** Aquaculture systems for wastewater treatment, in Seminar Proceedings and Engineering Assessment, 430/9-80-006, U.S. Environmental Protection Agency, Washington, D.C., 1979, 1.
21. **Tchobanoglous, G., Stowell, P., Ludwig, R., Coltm, J., and Knight, A.,** The Use of Aquatic Plants and Animals for the Treatment of Wastewater: An Overview, Departments of Civil Engineering and Land, Air, and Water Resources, University of California, Davis, 1980.
22. **Wolverton, B. C.,** Engineering design data for small vascular aquatic plant wastewater treatment systems, in Aquaculture Systems for Wastewater Treatment, EPA 430/0-80-006, U.S. Environmental Protection Agency, Washington, D.C., 1979, 179.
23. **Reddy, K. R.,** Nutrient transformations in aquatic macrophyte filters used for water purification, in Proc. 3rd Water Reuse Symp., San Diego, August 26, 1984, 660.
24. **Reddy, K. R. and Debusk, W. F.,** Growth characteristics of aquatic macrophyte cultures in nutrient enriched water: water hyacinth, water lettuce, and pennywort, *Econ. Bot.*, p. 225, 1984.
25. **Prokopovich, N., Storm, A., and Tennis, C.,** Toxic Trace Metals in Water Hyacinth in the Sacramento-San Joaquin Delta, California, U.S. Bureau of Reclamation, Sacramento, 1986.
26. **Wolverton, B. C. and McDonald, R. C.,** Water Hyacinths for Removing Chemical and Photographic Pollutants, Tech. Memo. x-72731, National Aeronautical and Space Administration, Washington, D.C., 1976.
27. **Stumm, W. and O'Melia, C. R.,** Stoichiometry of coagulation, *J. Am. Water Works Assoc.*, 60, 514, 1968.
28. **Stumm, W. and Morgan, J. J.,** Chemical aspects of coagulation, *J. Am. Water Works Assoc.*, Vol. 54, No. 8, 1962.
29. **Joyce, R. S., Allen, J. B., and Sukenik, V. A.,** Treatment of municipal wastewater in packed activated carbon beds, *J. Water Pollut. Control Fed.*, 38(5), 813, 1966.
30. **U.S. Environmental Protection Agency,** Manual on Activated Carbon Adsorption, U.S. EPA, Washington, D.C.
31. **Osborn, G. E.,** *Synthetic Ion Exchangers*, Chapman and Hall, London, 1961.
32. **Paulson, C. J.,** Chromate recovery by ion exchange, *Proc. Ind. Waste Conf.*, 36(6), 209, 1952.
33. **Cartwright, P. S.,** Innovative technology to treat toxic wastes from a thin film head manufacturing facility — a case history, in Proc. 3rd Water Reuse Symp., San Diego, August 1984, 778.
34. **Anon.,** Ozone counters waste cyanide's lethal punch, *Chem. Eng.*, 65, 63, 1958.
35. **Borchardt, J. A.,** The use of chlorine for oxidation of impurities in water, in Proc. Symp. Oxidation and Adsorption, Department of Civil Engineering, University of Michigan, Ann Arbor, 1965.
36. **Ladbury, J. W. and Cullis, C. F.,** Kinetics and mechanism of oxidation by permanganate, *Chem. Rev.*, 58, 403, 1958.
37. **Dee, N. et al.,** Environmental Evaluation System for Water Resources Planning, Final Report, Battelle-Columbus Laboratories, Columbus, OH, 1972.
38. **Eckenrode, R. T.,** Weighing multiple criteria, *Manage. Sci.*, 12(3), 180, 1965.
39. **Brown, R. L. and Beck, L. A.,** Subsurface agricultural drainage in California's San Joaquin Valley, in *Biotreatment of Agricultural Wastewater*, Huntley, M. E., Ed., CRC Press, Boca Raton, FL, 1989, chap. 1.
40. **Oswald, W. J., Chen, P. H., Gerhardt, M. B., Green, F. B., Nurdogan, Y., Von Hippel, D. F., Newman, R. D., Shown, L., and Tam, C. S.,** The role of microalgae in removal of selanate from subsurface tile drainage, in *Biotreatment of Agricultural Wastewater*, Huntley, M. E., Ed., CRC Press, Boca Raton, FL, 1989, chap. 9.

Chapter 3

CURRENT OPTIONS IN TREATMENT OF AGRICULTURAL
DRAINAGE WASTEWATER

Edwin W. Lee

TABLE OF CONTENTS

I. INTRODUCTION

The treatment of agricultural drainage is a technology research and development issue. It is a unique technical challenge since a review of the literature has not revealed any real experience in the management of large volumes of agricultural drainage containing high salt loads, including toxic trace elements. Therefore, a new technology must be developed, starting with basic laboratory research and followed by pilot testing of findings in a developmental process leading to practical field applications, in solving the problems of agricultural drainage. Conventional wastewater treatment technology cannot effectively meet the stringent requirements for the removal of toxic substances in accordance with proposed standards for receiving water standards and at costs affordable to the agricultural economy.

Salinity and its effects on agricultural lands have been recognized problems since the introduction of irrigated farming by man in ancient times.[1] History has recorded many incidences of fertile lands that have subsequently become barren due to salt. Salinity in irrigated agricultural lands of the San Joaquin Valley of California was recognized as a problem at the early planning stages of the Central Valley Project of the U.S. Bureau of Reclamation. Thus, drainage management as a salinity control measure was an important factor in the development plans. However, with time, the management issues presented changing obstacles as the simple approaches in salinity control of irrigated lands evolved into complex problems of water quality and environmental objectives. Thus, the traditional agricultural irrigation management required to sustain and optimize productivity by salinity control are made complex by the need to protect the environment from the impact of drainage. In the San Joaquin Valley, emphasis was shifted in the 1960s from traditional concerns with salinity to include the broader issues of eutrophication and contamination by pesticides. Recent discoveries of dead and deformed birds at Kesterson Reservoir,[2] a receiving basin for agricultural drainage in the central San Joaquin Valley, have focused attention on toxic trace elements, particularly selenium.

A comprehensive plan is currently under development for the management of agricultural drainage in the San Joaquin Valley. The San Joaquin Valley Drainage Program (SJVDP),[3] a state and federal interagency program, is developing the management plan. This chapter presents the current research efforts by scientists seeking options under the SJVDP for the treatment of agricultural drainage wastewater as part of the comprehensive plan for the management of the natural resources of the San Joaquin Valley. It is recognized that agricultural drainage problems in other locations in the U.S. and elsewhere may be different because of geology, hydrology, climate, and other factors. Nevertheless, the California experiences, particularly regarding residual nitrates, selenium, and other trace elements in drainage water, may have application in other agricultural areas.

II. THE APPROACH

The ultimate objective of research in agricultural drainage treatment technology is to develop methods and processes that can meet effluent limitations and receiving water standards so that beneficial uses can be maintained, in compliance with requirements of regulatory agencies. Because of the limited experiences on the subject to date, the strategy is to consider any treatment method that has potential to successfully meet this objective. Accordingly, various technologies have been screened, evaluated, and compared as to technical effectiveness and limitations, affordable costs, and environmental consequences.

In the development process, a phased approach was adopted, with selected decision points in which study results were analyzed before the next development stage of the technology would be undertaken. This approach involved the following sequential steps:

TABLE 1
Elements of Concern

Arsenic (As)	Manganese (Mn)
Boron (B)	Mercury (Hg)
Cadmium (Cd)	Molybdenum (Mo)
Chromium (Cr)	Nickel (Ni)
Lead (Pb)	Selenium (Se)
	Silver (Ag)

Note: Trace elements in San Joaquin Valley drainage water which are designated targets for development of control technology are listed.

1. Screen and select promising treatment methods.
2. Conduct laboratory evaluation studies (batch reactors).
3. Assess effectiveness.
4. Design and construct a pilot-scale plant (continuous flow).
5. Conduct a pilot-scale test.
6. Perform an engineering evaluation.
7. Assess the treatment effectiveness, economic costs, and environmental impacts of the process.

III. THE TARGETS

Early management studies of agricultural drainage wastewater focused on maintaining salt balance and reducing boron content, which were the traditional requirements for sustaining the agricultural productivity of the land. In subsequent years, agricultural drainage laden with residual nutrients from fertilizer applications led to concern for the potential eutrophication of receiving waters. By 1984, the targets of treatment technology were the trace elements, particularly those with toxic effects and the potential for accumulation and biomagnification in plants, fish, and wildlife. The list of trace elements of concern in the San Joaquin Valley is shown in Table 1. This list is based on a consideration of environmental criteria, as well as limitations of available monitoring data, and reflects the underlying need for protection of the beneficial uses of the land and water resources of the basin.

From a regulatory approach, targets are encompassed in environmental control programs. In the San Joaquin Valley, the state regulatory agency developed a basin plan for the protection of beneficial uses as early as 1965. The latest plan was approved for implementation by the California State Water Resources Control Board (SWRCB) in 1975. With the rising concern for toxic trace elements in the valley, the California SWRCB[4] was directed in 1985 to prepare revisions to the 1975 plan to take into account the discharge of agricultural drainage. This regulatory charge was unique in that nonpoint discharges of agricultural drainage water had never been an object of control actions in California.

In 1987, the California SWRCB prepared a plan[5] for regulating agricultural drainage in the San Joaquin River Basin. The water quality objectives for the San Joaquin River downstream of Hills Ferry are shown in Table 2; these represent the present targets for the development of a feasible treatment technology. Only four water quality parameters are under evaluation for control in this plan, including only three of the trace elements listed in Table 1. A broader list of trace element parameters may be developed later. Therefore, at this time there are uncertainties regarding the limitations on many other trace elements. It is likely that arsenic, cadmium, chromium, and mercury could be included in a long-term control program. Moreover, the SWRCB provided evidence in their background papers to support a reduction of selenium concentrations to 2 ppb in the river as a long-term objective.

TABLE 2
**Recommended Water Quality
Objectives for the San Joaquin
River**[5]

Selenium	5 ppb
Electroconductivity	1.0 mmho
Boron	700 ppb
Molybdenum	10 ppb

TABLE 3
**Potential Treatment
Technology for Agricultural
Drainage**

Desalination
  Thermal distillation
  Reverse osmosis
  Electrodialysis
  Freeze crystallization
  Vapor compression
Biological
  Bacterial
  Fungal
  Microalgal microbial
  Aquatic systems
  Silviculture
Chemical precipitation
  Iron hydroxides
  Alum
  Lime soda
  Sulfides
Adsorption
  Iron filing
  Ion-exchange resins
  Activated carbon
  Activated alumina
Electrochemical
  Electrocoagulation
  Electron-exchange resins

These uncertainties should be kept in mind in the development of any treatment technology for agricultural drainage wastewater in the San Joaquin Valley.

IV. POTENTIAL TECHNOLOGY OPTIONS

Numerous methods for treatment of municipal and industrial water and wastewater have been screened for possible application to agricultural drainage. A list of technologies that potentially could be applied to drainage water is shown in Table 3. This list was screened to select treatment methods with potential for further development.

To provide a comprehensive approach to the management of agricultural drainage, the plan included efforts to limit and control the emission of wastewater and pollutant loads from irrigated agricultural lands, on-farm treatment methods, regional pollution control facilities, and, finally, disposal and reuse methodologies. The application of a holistic approach was considered necessary to attack the enormous problem posed by agricultural drainage water in irrigation areas covering more than 2 million acres in the San Joaquin Valley.

TABLE 4
Drainage Water Analysis — San Luis Drain at Mendota, CA

Constituent	Unit	Average[31]	Maximum[31]
Sodium	mg/l	2,230	2,820
Potassium	mg/l	6	12
Calcium	mg/l	554	714
Magnesium	mg/l	270	326
Alkalinity	mg/l $CaCO_3$	196	213
Sulfate	mg/l	4,730	6,500
Chloride	mg/l	1,480	2,000
Nitrate/nitrite	mg/l	48	60
Silica	mg/l	37	48
TDS	mg/l	9,820	11,600
Suspended solids	mg/l	11	20
Total organic carbon	mg/l	10.2	16
COD[a]	mg/l	32	80
BOD[b]	mg/l	3.2	5.8
Temperature[c]	°C	19	29
pH	—	8.2	8.7
Boron	μg/l	14,400	18,000
Selenium	μg/l	325	420
Strontium	μg/l	6,400	7,200
Iron	μg/l	110	210
Aluminum	μg/l	<1	<1
Arsenic	μg/l	1	1
Cadmium	μg/l	<1	20
Chromium (total)	μg/l	19	30
Copper	μg/l	4	5
Lead	μg/l	3	6
Manganese	μg/l	25	50
Mercury	μg/l	<0.1	<0.2
Nickel	μg/l	14	26
Silver	μg/l	<1	<1
Zinc	μg/l	33	240

Note: mg/l = parts per million by weight (ppm); μg/l = parts per billion by weight (ppb).

[a] Chemical oxygen demand.
[b] Biological oxygen demand.
[c] Temperature varied from between 23 and 25°C (summer) to between 12 and 15°C (winter).

An initial review of available treatment methods that are known to engineering practitioners for municipal and industrial wastewater revealed many difficulties in application to agricultural drainage, mainly because of the complex nature of chemical constituents. A typical chemical analysis of drainage water taken from agricultural drains is shown in Table 4. The sampling site is located on the San Luis Drain, which drains over 40,000 acres of irrigated lands in the west side of the San Joaquin Valley and represents a typical mix of drainage water.

Several characteristics of this drainage wastewater should be noted because they contribute to the complexity of treatment. The average total dissolved solids (TDS) is 9820 mg/l, which is an indication of heavy salt content. The sulfate (SO_4) content is about 472 mg/l, in contrast to selenium (0.325 μg/l). As these two chemicals are similar, the overabundance of sulfates dominates any method for removing selenium. Selenium exists primarily in the selenate form (~90%), with the remainder (~10%) existing as selenite. The selenate form, which is the highest oxidation state, is considered the most difficult to reduce and to remove from solution. Because of these complex chemical characteristics, no easy and simple process is known for the reduction of selenate and other trace elements in drainage water. Furthermore, disposal and reuse are equally complicated by the salt load, even after the removal of trace elements. The feasible solutions are probably few in number when technical effectiveness and economic affordability are considered.

V. TREATMENT OPTIONS UNDER DEVELOPMENT

Requirements for research and development of a treatment technology for the removal of toxic trace elements in agricultural drainage water exist today only in relation to the problems in the San Joaquin Valley. In the industrial wastewater field, limited research has been conducted for selenium removal, mainly in relation to uranium mining wastes and coal ash ponds. Technology for the removal of trace toxic substances has been developed for drinking water supplies, which generally have much lower salt content and are chemically less complex than agricultural drainage.

Earlier studies in agricultural drainage water were directed at the control of salts and boron (for the protection of agriculture) and at the reduction of nitrogen (which was identified as a potential cause for the eutrophication of receiving waters). Current studies in the San Joaquin Valley are focused on salinity and selenium, since these two substances present the most difficult technological challenge. Boron and other trace elements have been identified by the SWRCB for regulatory action, and other trace elements have been identified as cause for concern. However, these trace elements do not present the technical difficulties and economic constraints expected from salt removal and selenium reduction. It is believed that any removal process which can be successfully applied to salts and selenium can also remove other trace elements at the same time. The level of removal will have to be studied as part of the comprehensive treatment program.

At this stage of the program, an effective technology has not been developed to determine the feasibility for salt and selenium reduction, although several promising processes (described in the following sections) are under evaluation. In screening the various treatment methods shown in Table 3, many were dropped from further consideration because of technical ineffectiveness or high cost. The following presentation is a review of treatment options currently under research and development.

The technologies selected for the treatment of agricultural drainage water for salt and selenium removal can be broadly categorized into five basic systems. Although there may be a mixture of systems in process streams, these are noted separately as follows:

- Desalination
- Biological
- Chemical precipitation
- Adsorption
- Electrochemical

A. DESALINATION

Desalination is a well-developed technology which has been employed at many locations in the U.S. and worldwide. The most promising method selected for agricultural drainage investigation is reverse osmosis (RO). This technology has been used to convert seawater to drinking water at an affordable cost at locations where water is scarce or for specialized objectives. Other desalination technology was not considered cost effective for agricultural drainage treatment and consequently was dropped from further consideration. The current program involves two independent studies, both of which consider desalination technology and the removal of trace elements, including selenium. The first is a feasibility study, completed under a contract to CH2M-Hill,[6] to consider the applicability of RO. The second is a pilot field experiment, still being conducted by the California Department of Water Resources (DWR). The CH2M-Hill study assumed the use of off-the-shelf technology. Because of the complexity of chemical constituents in drainage water, pretreatment was considered necessary to maintain optimum salt rejection capability for RO systems. Pretreatment technology involved chemical stabilization of RO feed water and was a major cost component in the process. Using a three-stage RO system with chemical pretreatment, the consultant estimated costs at around $1000/acre-ft (U.S., 1985) to reduce TDS to 550 to 650 ppm and selenium to below 10 to 20 μg/l. Product recovery (salt harvesting, desalted water, reuse, and conjunctive solar ponds) may reduce these high costs, but the economics of off-the-shelf RO process do not provide any note of optimism for agricultural drainwater at this time.

At the Los Banos (California) Desalting Facility, the DWR has been testing RO technology at a pilot plant; drainage water is being desalted for reuse and the reject brine used for electric power generation in solar ponds. Pretreatment facilities have included biological wetlands and chemical precipitation to stabilize the RO feed water. Marinas and Selleck[7] assessed the operation of these RO units in 1985. The TDS was reported to be reduced from >9000 mg/l to about 145 mg/l and selenium from 266 μg/l to about 1 μg/l. Related costs were not indicated, but a report on these studies is scheduled to be released by the DWR in 1990.

B. BIOLOGICAL

One of the most promising biological treatment processes under development involves anaerobic bacteria in enclosed reactors. Treatment of agricultural drainage water with anaerobic bacteria in upflow filters and deep ponds for the purpose of reducing eutrophic potential of discharges into the delta was studied in the 1960s by the California DWR[8] at Firebaugh. More recently, Binnie California, Inc. of Fresno, in cooperation with local agricultural interests, the Westlands Water District, and the DWR, has been studying this process for the removal of selenium. Work has been underway since 1985 at a small-scale pilot plant at Murietta Farms, near Mendota, CA.[9]

The methodology (Figure 1) consists of processing water through a series of biological reactors, microfilters, and, as a final polishing step, through ion-exchange resins. A carbon source is injected into the influent flow entering the anaerobic reactors. Typically, methanol is used as the degradable carbon source for bacterial energy and growth. Typical detention time (hydraulic) in the biological reactors is 1.5 to 2.5 h, but there is no information on the detention time of the bacterial biomass. It was reported by Binnie California, Inc.[9] that nitrate in the influent drainage water (~150 mg/l) must be reduced before selenate can be removed. Influent selenium concentrations of 350 to 500 μg/l have been reported to be reduced in the biological reactors to approximately 40 to 50 μg/l. Effluent is then filtered through a crossflow microfilter constructed of a fine fabric. These filters are capable of removing particles as small as 0.1 μm in diameter. A precoat layer is applied to the fabric material, and retained solids are moved downstream and out of the system by continuous flow and periodic purging. Filtered water is passed through the fabric. The filtration step further reduces selenium to 10 to 20 μg/l. The final step in the treatment process involves ion-exchange resins to remove boron, which also can reduce the remaining selenium to <10 μg/l.

A detailed report on these studies was prepared by Binnie California, Inc. and released in 1988.[31] Costs are tentatively reported to be about $150/acre-ft, based on data from the pilot plant. To obtain engineering design parameters and firm cost data, a large pilot plant (500,000 gal/d) is currently under design. This plant was scheduled for operation in 1988, but the plan has been put on hold.

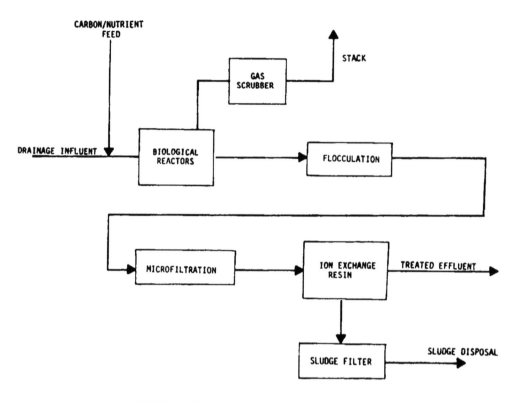

FIGURE 1.   Bacterial selenium removal plant schematic.

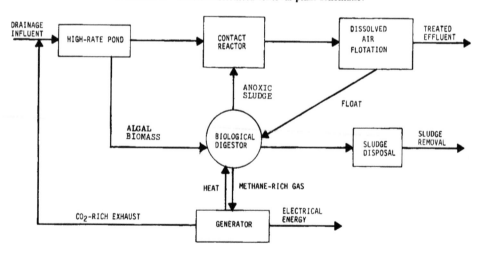

FIGURE 2.   Microalgal bacterial selenium removal plant schematic.

Another biological method under study, but still at an early stage of development, involves a unique microalgal bacterial process. The proposed system[10] combines a high-rate algal pond with an algal digesting process. Its primary elements are shown in Figure 2. The high-rate algal pond is fed with drainage water for photosynthetic biomass growth. In a separation chamber, algae undergo bioflocculation and then removal by gravity or dissolved air flotation. The biomass is removed for digestion and the production of methane, which can be used as a fuel source. The carbon dioxide by-product of fuel combustion is fed to the high-rate pond as a carbon source for algal growth.

High-rate algal ponds can produce up to 300 mg/l of algae (dry weight) on agricultural drainage water. The selenium content of algae grown on drainwater is about 85 mg Se per kilogram of dry algae. Thus, the algal biomass can remove a maximum of only 25 μg/l in drainage water. However, the digestor effluent is a strong reducing agent and, when added to the high-rate algal pond effluent, will reduce selenium to insoluble forms. Laboratory results[11] have shown that selenium can be reduced to 1 μg/l by digestor solids. At this early stage of process development, treatment costs are not firm.

Another microbial process, reported by the U.S. Bureau of Reclamation,[12] involves the bacterial reduction of selenium to insoluble forms in the upper layers of the bottom sediments of Kesterson Reservoir. The immobilization of selenium by the process would prevent migration into the groundwater system under the reservoir. This was proposed by Weres et al.[13] as a cleanup process for Kesterson Reservoir, and joint studies by the U.S. Bureau of Reclamation and the University of California have been underway since 1985. However, in 1987 the California SWRCB directed that the cleanup of Kesterson should be accomplished by removal and stockpiling of the contaminated sediments into containment areas.[4]

Based on laboratory observations of the volatilization of selenium by fungi, Frankenberger[14] proposed studies for the reduction of selenium from the sediments of Kesterson Reservoir and other highly seleniferous areas on the west side of the San Joaquin Valley. This fungal process is accelerated by the application of an organic nutrient source and additional micronutrients. The microbial pathways and related kinetics of volatilization were studied by Frankenberger. Mass balances were evaluated and selenium forms studied, including the toxicity potential of the volatilized products. His findings were reported in 1988.[32]

Frankenberger[15] also proposed to the SJVDP a microbial process involving the use of fungi to volatilize selenium from evaporation ponds, which are currently used for disposal of drainage water in the San Joaquin Valley. The process, which is similar to that proposed for selenium volatilization from soil and sediments,[14] would also require organic energy sources which, in this case, might be provided by the decomposition of stimulated algal growth. This process is currently under study; laboratory studies were initiated in late 1987, and an initial report came out in 1988.[33]

Other investigators have proposed the use of aerobic bacteria to reduce selenium to easily removed metalloid forms. Gersberg et al.[16] of the San Diego Regional Water Reclamation Agency conducted laboratory-scale studies using aerobic bacteria and reported on the detoxification of selenate. Altringer[17] is studying the use of rotary biological contactors for the removal of selenium. Although this is also basically an aerobic process, the biological film on the contactor may be near anoxic conditions below the surface layers.

C. PRECIPITATION

Chemical precipitation of trace elements in wastewater is a well-developed engineering technology, particularly in the treatment of industrial wastes for the removal of heavy metals. The application of this technology to selenium has only been successful in the reduction of selenite. Sorg and Logsdon[18] used ferric sulfate to reduce selenite from 30 to 10 μg/l in drinking water supplies. Merrill et al.[19] performed pilot plant testing on coal ash pond effluent using iron oxyhydroxide and removed more than 80% of the selenite, but only about 10% of the selenate. The selenate form is not easily susceptible to chemical precipitation and, since selenate is the dominant form (90%) of selenium in agricultural drainage water in the west side of the San Joaquin Valley, chemical treatment processes have not been fully developed for successful application to date.

The U.S. Bureau of Reclamation Engineering and Research Center (E & R) in Denver, CO has completed initial laboratory studies on the use of ferrous hydroxide for the reduction and precipitation of selenium from agricultural drainage water. Initial studies[20] indicate that selenium could be reduced from 250 to approximately 1 μg/l, an exceedingly high removal

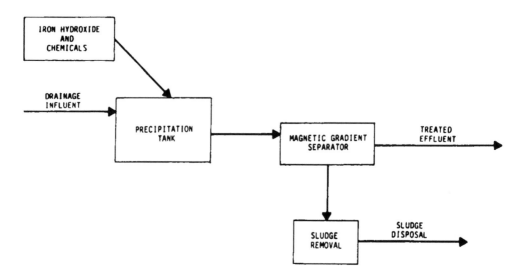

FIGURE 3.   Iron hydroxide treatment selenium removal plant schematic.

performance. Large-scale batch reactor studies are currently underway at the E & R laboratory and in the west side of the San Joaquin Valley. An engineering feasibility study based on these studies will be prepared to determine removal rates, chemical reaction kinetics, and preliminary costs. If the engineering evaluations are promising, recommendations will be made to continue studies in a continuous-flow, small-scale pilot plant. The design, construction, and operation of this pilot plant will provide technical data and information on the performance of the chemical precipitation process with respect to technical effectiveness for removal of selenium and other trace elements, cost economics, and environmental acceptability of the process stream and waste by-products. A schematic diagram of the batch chemical precipitation process as tentatively proposed is shown in Figure 3.

D. ADSORPTION

The adsorption of selenium on activated alumina has been reported by Sorg and Logsdon,[18] who indicated that treatment resulted in 95% removal of selenite and selenate in spiked distilled water. However, selenite reduction fell to 62 to 68% and no selenate was removed from a hard groundwater. It was thought that sulfate competition reduced adsorption rates. Trussell et al.,[21] also using activated alumina, showed that selenite removal was favored over selenate.

Baldwin et al.[22] of the Kerr-McGee Corporation (Oklahoma City) patented a process for the removal of selenium with iron filings. Selenate was reduced to selenite and selenium in a solution with a pH range of 3.0 to 5.5. The process was used to treat uranium mining waste for the removal of selenate. The selenate either formed a precipitate or combined with the iron metal.

Mayenkar,[23] of the Harza Engineering Company, Chicago, patented an adsorption process for the removal of heavy metals with iron filings. It was claimed that the surface of the iron filings could be activated by oxygenation and that selenium was adsorbed on the activated surface. Removal of iron, nickel, zinc, copper, lead, cadmium, mercury, and arsenic was reported. This treatment process was tested with agricultural drainage water at the Panoche Drainage District in the the San Joaquin River Basin in 1986. An initial report was prepared by Harza Engineering[24] on their pilot plant tests performed under field conditions and in a continuous-flow regime. The range of removal at different overflow rates and contact times varied from 20 to 90% of total selenium. The process appears to be similar to the previously described Kerr-McGee process,[22] except that the pH was not adjusted in the Harza field tests.

In the Harza process, the removal of selenium was greatly influenced by the overflow rate. A major problem encountered in both upflow and downflow columns was the cementation of the iron filings. This required mechanical loosening of the media to maintain flow. The cause of plugging of the column was not explained, but possibly could be due to precipitation or to a combination of agricultural drainwater constituents with the iron filings. Harza Engineering is planning to study the iron filing adsorption process in conjunction with the SJVDP to determine if precipitation or chemical bonding of constituents is involved. Meanwhile, the Panoche Drainage District is planning the construction in 1989 of a large-scale pilot plant to study the iron filing process with Harza Engineering.

The removal of trace elements with selective ion-exchange resins is a theoretical possibility, as proposed by Herrmann.[26] Maneval et al.[27] reported the effective removal of selenate from drinking water. This method is currently under study by the Bakersfield, CA office of Boyle Engineering Corporation[28] and the SJVDP for agricultural drainage wastewater. Initial bench-scale studies have indicated selectivity for selenium exchange, and these characteristics will be exploited in further studies to develop a specific resin for application to drainage wastewater.

Sposito[29] proposed a study of the attenuation of selenium from irrigated agricultural soils with *in situ* chemical amendments. Preliminary laboratory studies with soils on the west side of the San Joaquin Valley indicated this to be a promising process capable of reducing the emission of selenium loads from farms. Laboratory studies of this process were initiated in 1987, with funding support from the SJVDP.

E. ELECTROCHEMICAL

The U.S. Bureau of Reclamation's E & R Center studied electron-exchange resins,[20] which are capable of acting as catalysts for the reduction of selenate and selenite. Hydrazine, hydroquinone, and hydroxyozine are possible active groups for electron exchange. Initial studies with an old batch of resins showed promise, but when a new batch was purchased from the original manufacturer the results were less indicative of an electron-exchange potential, and studies were terminated.

Another electrochemical process involves the application of an electrical field across sacrificial electrodes immersed in drainage water to precipitate and remove trace elements. Andco Environmental Processes, Inc. of Amherst, NY, in 1985 field tests of a small-scale electrocoagulation plant near Mendota, was able to remove up to 50% of the selenium from drainage water in a single-pass system.[30] Further studies are needed to determine if a multiple-pass process could increase the effectiveness of selenium removal and to obtain engineering operation and cost data to evaluate the feasibility of such a process.

F. OTHER OPTIONS UNDER REVIEW

The prior sections have identified options currently under development and represent a broad approach to the problem, supported with study efforts currently underway. There are other options with some degree of promise that have been under review, but which have not been followed up with detailed studies to date. These include the use of aquatic plants, aquaculture (fishery and shellfish), and silviculture systems, all of which are primarily for the disposal and reuse of drainage waters, although treatment by biological uptake and removal is inherent in these processes.

VI. SUMMARY

At this date, the real options for treatment of agricultural drainage wastewater are limited. While the removal of salts is a well-developed technology, the reduction of trace elements, particularly selenium, presents major developmental problems. Several processes are now under development, and promising results from laboratory and small-scale pilot plant studies have been reported. The development of these few processes to feasibility scale presents a major technical challenge which must be overcome for the protection of the environment of the San Joaquin Valley.

REFERENCES

1. **Food Agriculture Organization,** Irrigation Drainage and Salinity, Food Agricultural Organization, United Nations, Rome, 1973.
2. **Ohlendorf, H. M., Hoffman, D. J., and Aldrich, T.,** The biologic system, presented at the Conf. Toxicity Problems at Kesterson Reservoir, California, Sacramento, December 5 to 7, 1983.
3. **San Joaquin Valley Drainage Program,** Developing Options — An Overview of Efforts to Solve Agricultural Drainage and Related Problems in the San Joaquin Valley, SJVDP, Sacramento, 1987.
4. **California State Water Resources Control Board,** Sacramento, Order No. 85-1, February 1985, and Order No. 87-3, March 1987.
5. **California State Water Resources Control Board,** Regulation of Agricultural Drainage to the San Joaquin River, Tech. Comm. Rep., SWRCB, Sacramento, 1987.
6. **CH2M-Hill,** Reverse Osmosis Desalting of the San Luis Drain Conceptual Level Study, prepared for SJVDP under U.S. Bureau of Reclamation contract, CH2M-Hill, Emeryville, CA, 1986.
7. **Marinas, B. J. and Selleck, R. E.,** The Removal of Selenium with Reverse Osmosis, in *Proc. 1st Annu. Environ. Symp. Selenium in the Environment,* Publ. CAT1/860201, California Agricultural Technology Institute, California State University, Fresno, 1986.
8. **California Department of Water Resources,** Denitrification by Anaerobic Filters and Ponds, Phase II, USEPA Rep. 13030, ELY-06/71-14, DWR, Sacramento, July 1971.
9. **Binnie California, Inc.,** Prototype Selenium Removal Plant, Report to Westlands Water District, California, Binnie California, Inc., Fresno, 1987.
10. **Oswald, W. J.,** Potential for Treatment of Agricultural Drain Water with Microalgal Bacterial Systems, prepared for SJVDP under U.S. Bureau of Reclamation contract, University of California, Berkeley, 1985.
11. **Oswald, W. J.,** Microalgal Bacterial Treatment for Selenium Removal from San Joaquin Valley Drainage Waters, prepared for U.S. Bureau of Reclamation and California Department of Water Resources, University of California, Berkeley, 1987.
12. **U.S. Bureau of Reclamation,** Final Environmental Impact Statement, Kesterson Program, U.S. Bureau of Reclamation, Sacramento, 1986.
13. **Weres, O., White, A. F., Wollenberg, A. A., and Gee, A.,** Geochemistry of Selenium at Kesterson Reservoir: Possible Remedial Measures, Earth Sciences Department, Lawrence Berkeley Laboratory, University of California, Berkeley, 1985.
14. **Frankenberger, W.,** In-Situ Volatilization of Selenium, Cooperative Agreement, University of California-Riverside and U.S. Bureau of Reclamation, Sacramento, 1987.
15. **Frankenberger, W.,** In-Situ Volatilization of Selenium: Evaporation Ponds, Cooperative Agreement, University of California-Riverside and U.S. Bureau of Reclamation, Sacramento, 1987.
16. **Gersberg, R. M., Brenner, R., and Elkins, B. V.,** Removal of Selenium Using Bacteria, in *Proc. 1st Annu. Environ. Symp. Selenium in the Environment,* Publ. CAT1/860201, California Agricultural Technology Institute, California State University, Fresno, 1986.
17. **Altringer, P. B.,** A Biohydrometallurgical Approach to Selenium Removal, U.S. Bureau of Mines, Salt Lake City, 1987.
18. **Sorg, T. J. and Longsdon, G. S.,** Treatment technology to meet interim primary drinking water regulations for inorganics, *J. Am. Water Works Assoc.,* 70, 379, 1978.
19. **Merrill, D. J., Manzione, M. A., Peterson, J. J., Parke, D. S., Chow, W., and Hobbs, A. O.,** Field evaluation of arsenic and selenium removal by iron coprecipitation, *J. Water Pollut. Control Fed.,* 58, 18, 1986.
20. **U.S. Bureau of Reclamation,** Selenium Removal Studies, Applied Sciences Referral Memorandum to Mid-Pacific Region, E & R Center, USBR, Denver, 1986.
21. **Trussell, R. R., Trussell, A., and Kreft, P.,** Selenium Removal from Ground Water Using Activated Alumina, EPA-600/2-80-153, Municipal Engineering Research Laboratory, U.S. Environmental Protection Agency, Cincinnati, 1980.
22. **Baldwin, R. A., Stanter, J. C., and Terrell, D. L.,** Process for Removal of Selenium From Aqueous Systems, U.S. Patent 4,405,464, 1983.

23. **Mayenkar, K. V., Lagvankar, A., and Pherson, P. A.,** Innovative process for removing heavy metals from wastewater, presented at Annu. Meet. Central States Water Pollution Control Assoc., May 14 to 16, 1986.

24. **Harza Engineering,** Selenium Removal Study, prepared for Panoche Drainage District, California, Harza Engineering, Chicago, 1986.

25. **Harza Engineering,** Iron Filing Research, prepared with San Joaquin Valley Drainage Program under U.S. Bureau of Reclamation contract, Harza Engineering, Chicago, 1987.

26. **Herrmann, C. C.,** Removal of Ionic Selenium from Water by Ion Exchange, Review of Literature and Brief Analyses, report to U.S. Bureau of Reclamation, Water Thermal and Chemical Technology Center, University of California, Berkeley, 1985.

27. **Maneval, J. E., Sinkovic, J., and Klein, G.,** Selenium Removal from Drinking Water by Ion Exchange, Rep. CR-810254-01-1 to U.S. Environmental Protection Agency, Water Thermal and Chemical Technology Center, University of California, Berkeley, 1984.

28. **Boyle Engineering Corporation,** Selenium Selectivity in Ion Exchange Resins, prepared for SJVDP under U.S. Bureau of Reclamation contract, Boyle Engineering, Bakersfield, CA, 1987.

29. **Sposito, G.,** Attenuation of Selenium in Irrigated Agricultural Soils, cooperative agreement between University of California, Riverside and U.S. Bureau of Reclamation, Sacramento, 1987.

30. **Andco Environmental Processes, Inc.,** Andco Electrochemical Selenium Removal Process, Andco Environmental Processes, Inc., Amherst, NY, 1985.

31. **Binnie California, Inc. and California Department of Water Resources,** Performance Evaluation of Research Pilot Plant for Selenium Removal, Binnie California, Fresno, 1988.

32. **Frankenberger, W.,** Dissipation of Soil Selenium by Microbial Volatilization at Kesterson Reservoir, final report for U.S. Bureau of Reclamation, Sacramento, CA, 1988.

33. **Frankenberger, W.,** In-Situ Volatilization of Selenium: Evaporation Ponds, report to San Joaquin Valley Drainage Program, Sacramento, CA, 1988.

Chapter 4

BIOTECHNOLOGY IN ENVIRONMENTAL ENGINEERING

Gedaliah Shelef

TABLE OF CONTENTS

I. INTRODUCTION

One of the principal tasks of the environmental engineering profession (formerly sanitary engineering) is the treatment of liquid, semiliquid, and solid organic wastes of municipal, agricultural, and industrial origin. The prime "raison d'être" of this profession, i.e., the prevention of large-scale waterborne epidemics which had rampantly threatened the civilized world until the first quarter of this century, has been replaced by prevention of damage to the environment and to health by various substances, most of them organic, but many of them refractory, recalcitrant, xenobiotic compounds which are difficult to treat and remove.

Biological treatment using bioengineering processes (hence, biotechnological methods) has been established as the most efficient and economical way to remove organic compounds once a biological metabolic pathway exists. A brief attempt by some able chemical engineers, who joined the environmental engineering profession in the early 1960s, to replace mainstream biological processes by introducing physicochemical methodology proved to be unsuccessful. Since the early 1970s we have witnessed an upsurge of even more sophisticated and efficient biological processes, backed by process kinetics, contemporary biotechnological reactor design, and better process control. For example, when nitrogen removal from wastewater became an essential task as eutrophication of inland lakes became a major environmental problem, physicochemical ammonia stripping was introduced, but it very soon gave way to nitrification-denitrification biological processes.

The 1980s have been marked by new advances in biological treatment processes, rendering them even more efficient than before, with the advent of rotating disk aerobic treatment, upflow anaerobic sludge blanket, and nutrient stripping high-rate algal ponds, to name just a few. With the problem of hazardous wastes being highlighted in this decade, the need to apply biological processes to treat and remove difficult-to-degrade refractory and recalcitrant compounds, possibly assisted by the recent advances in recombinant DNA, gene cloning, and DNA probing technologies, is indeed a challenge of enormous proportions.

II. ENVIRONMENTAL ENGINEERING BIOTECHNOLOGY

In the treatment of wastes, environmental engineering biotechnology (EEB) has used almost every type of process bioengineered in a microorganism that "classical" fermentation biotechnology (CFB) ever used. These processes include (1) aerobic fermentation (both suspended and immobilized biomass), (2) anaerobic fermentation (mesophilic and thermophilic), (3) anoxic processes (such as denitrification), (4) alcohol fermentation of municipal and agricultural solid wastes, and (5) microalgal biomass processes.

Almost any type of reactor or reactor configuration in classical fermentation biotechnology has been used in EEB systems (including innovations yet to be adapted in the classical one). These include continuous steered tank reactors (CSTR, with or without recirculated biomass), multistage, batch, plugflow, fixed media, upflow, fluidized bed, sludge blanket, race tracks, rotating disks, and so on.

The whole range of physical consistency, rheology, and viscosity of the substrate is employed in EEB, ranging from solids to sludges to slurries and liquids, down to extremely dilute substrate concentrations (such as nitrate removal from groundwater by fixed media nitrification-denitrification processes).

As a matter of fact, one could argue that as far as complexity, challenge, sophistication, size, process control, and process kinetics are concerned, CFB, with its stainless steel axenic cultures and sterile substrate, is *no match* to EEB, with its mixed culture, heterogeneous substrates, although carried out in "simple" reinforced concrete or steel (and sometimes fiberglass or even PVC) reactors.

The main differences between EEB and CFB are discussed in the remainder of this section.

A. SUBSTRATE

EEB uses a heterogeneous substrate, sometimes measured by its general characteristics. Thus, organic substrate can be measured as biochemical oxygen demand (BOD), chemical oxygen demand (COD), total organic carbon (TOC), etc. Utilizing, removing, or degrading the organic substrate can involve separate, mixed, sequential, consecutive, or simultaneous processes, not necessarily carried out by the same microorganism in the heterogeneous "soup". These processes may include sorption, incorporation, deactivation, detoxification, and degradation.

On many occasions the substrate has a zero or even a negative value, which favorably affects the economic feasibility of the process. Thus, for example, the ethanolic fermentation of corn has to face the cost of the grain being between one half and two thirds of the overall process cost, while with ethanolic fermentation of municipal solid wastes, the tipping fee of the substrate could range from $12* to $45 per ton or even more. (Up to $900 per ton has recently been offered by hard-pressed municipalities.) On the other hand, the wasteborne substrate in many instances has to pass a pretreatment stage.

B. SUBSTRATE CONCENTRATION AND PROCESS KINETICS

Unlike CFB substrate, the substrate in EEB is usually the food and/or energy source for the microorganisms; thus, its higher concentration in the reactor should enhance the bioreaction rate. At the same time, however, the removal of the very same substrate should be maximal; hence, its concentration in the reactor outflow should be minimal in order to meet receiving water-body effluent requirements, for example. This imposes an intrinsic conflict as far as process kinetics, design, and control are concerned, which makes EEB processes much more challenging than CFB processes, where substrate concentration may be at any designed concentration that will maintain the optimal bioreaction rate for converting substrate into product.

C. BIOMASS

The origin of the wasteborne substrate, its flow (feedstock rate), and the size and construction of the reactors render sterilization or even pasteurization of the substrate in EEB processes impossible. Rare exceptions occur in dealing with industrial or agroindustrial wastes, where some waste streams are inadvertently sterilized or where pretreatment of the wastes renders them sterile. The high-pressure/temperature, low-acid, short-duration hydrolysis of cellulosic substances in municipal solid wastes into fermentable sugars prior to ethanol fermentation is one example of the latter process. The luxury, therefore, of working with "pure" microorganism species or strains in axenic cultures, cherished so much by CFB, is virtually unattainable in EEB.

One of the challenges of EEB is to find the optimal combination of process design and reactor "ecological" conditions to render a desired species or strain predominant. There are important advantages of a mixed culture over a pure one in constantly allowing selection of the most fit. Furthermore, the opportunity exists for another species or strain to "take over" when adverse conditions malaffect the previous predominant one, thus not "losing the culture" and gradually, smoothly (or even automatically) "filling the vacuum" (which in mixed cultures almost never occurs when substrate and energy sources are abundantly available) by the acclimatization of another species/strain omnipresent in the "soup" to become the "new predominant".

Obviously, under conditions of heterogeneous substrate, the mixed culture has the advantage of simultaneously or consecutively being able to treat components in the substrate "soup" by respectively adapted microorganisms in the biomass "soup". The single sludge

---

* All monetary values are in 1987 U.S. dollars.

— activated sludge — nitrification — denitrification — phosphorus removed sequence, all attained in one mixed culture (although posing as various stages in the reactor), is a good example.

D. SIZE, FLOW, AND PROCESS CONTROL

For daily waste flows of between 15,000 and 25,000 m³ of municipal wastewater per 100,000 inhabitants per day or between 100 and 250 tons of municipal solid waste per 100,000 inhabitants per day, given as examples, reactor sizes range from 5,000 to 35,000 m³ for a typical activated-sludge aerobic fermentation per 100,000 inhabitants to multiple anaerobic digestion reactors of 3,000 m³ each for sewage sludge or municipal solid waste. The reactor sizes and flow (or feedstock) rates in EEB, therefore, dwarf even the most sizable CFB reactors when cities of over 1 million inhabitants are considered.

The variations in climatic conditions and in hourly, daily, and seasonal flow rates, the possibility of toxic or inhibiting substances inadvertently introduced into the reactor, and the daily variations in substrate composition and concentration (shutdowns of contributing industries during weekends and holidays) all make process design and control a most challenging endeavor unparalleled in CFB with its predesigned uniform flow, constant concentration, and steady-state condition.

The time is therefore long overdue to recognize EEB as biotechnology at its best and to overcome its being "dirty biotechnology". For many years, EEB was considered a "bastard" by many "clean" biotechnologists. Both history and genetics indicate that some hybrid-vigorous "bastards" have proven to be stronger, predominant, and certainly more colorful than "clean", "pure" inbreds.

III. BIODEGRADABILITY AND BIOCONVERSION

A simplified "law of biodegradability" indicates that any organic substance that is a product of anabolism has a catabolic pathway (usually carried out by microorganisms) leading to its biodegradation. This is also a key to all natural biological cycles and, indeed, the key to cybernetic prevention of destructive depletion or accumulation.

If, indeed, this law constitutes an "infinite truth", the time scale, turnover, and degree may vary, however, by orders of magnitude. Thus, the biodegradation (or mineralization) of urea into ammonia and CO_2, for example, can be carried out within seconds or minutes and the biodegradation of lignin by white mold in months; tree resins can resist biodegradation for centuries and, after volatilization of their lighter fractions, even millions of years, as evidenced by Baltic amber estimated to be 60 to 80 million years old.

Naturally produced organic compounds might be rendered nonbiodegradable by extreme conditions of high temperature and/or pressure, by chemical "denaturalization", by physicochemical alterations, or even by long-term catalytic surface reactions (as believed to be the case in the conversion of plant or animal lipids into petroleum which, as organic matter and silts and clays, had coprecipitated some 500 million years ago). Natural "denaturation" of organic matter to produce coal, petroleum, and tar, as well as halogenation, catalytic production, and polymerization of organic compounds, are the main sources of nonbiodegradable organics.

Even without the aid of genetic engineering, the success in rendering nonbiodegradable compounds biodegradable and treatable by biological processes has been moderate in the past 20 years, mostly by patient adaptation, selection, and acclimatization of bacterial species and strains to difficult-to-treat organic wastes. Phenol-rich petroleum wastewaters are treated by acclimated activated sludge; crude oil emulsions with seawater in oil tanker ballast waters can be removed by bacteria, and DDT and other compounds can be rendered biodegradable by altering anaerobic and aerobic digestion. Our research group, for example, has selected and acclimated anaerobic bacteria to treat toluene- and xylene-rich wastes from the pesticide manufacturing industry. The bacteria were collected from a natural earth depression in a deep gorge where the raw wastes had been discharged for many years; the isolates were then painstakingly acclimated in a bench-scale reactor.

While halogenated organic compounds, particularly the aromatic ones (halogenated benzenes, phenol, furans, and dioxins), constitute a most difficult case, the combination of multiple biological processes using adapted microorganisms that dehalogenate, demethylate, deacetylate, hydroxylate, cleave, reduce, or oxidize, with or without physical, chemical, or physicochemical pretreatment, should yield better results.

Preferential sorption and luxury uptake of various organic and inorganic compounds by microorganisms has been quite successful, particularly by using microalgae in high-rate ponds. Unfortunately, the concentrated biomass in this case should be disposed of rather than used as a source of animal feed or fertilizer.

IV. THE QUEST FOR A ''SUPERBUG''

Since the introduction of biological wastewater treatment, EEB has been questing for a ''superbug'' which will either markedly increase process efficiency or degrade and remove difficult-to-treat compounds.

Selection, adaptation, acclimatization, and even induced mutation have been tried, and vendors have sold concentrated bacterial cultures of ''superbugs'' promising astonishing results. On at least two occasions, we followed the fate of such labeled ''superbugs'' in field-scale biological reactors to find, to our dismay, that they disappeared within a week. This happened after the efficiency of the selected strain in bench-scale reactors was found to be quite promising.

As far as selection is concerned, it is hard to believe that one can compete in the laboratory with the natural selection of the ''most fit'' in a mixed culture under field conditions.

Genetic engineering, with its new advances in recombinant DNA, gene cloning, and DNA probing, can be most instrumental in the ''creation'' of the long-sought ''superbug''. Assuming such a gene-altered bacterium will be introduced and tested in the laboratory, classical biotechnology will still have to determine the optimal ecological conditions necessary for this organism to grow, and they will have to find proper reactor design and reactor operating conditions which favor the new organism's capabilities.

The success of genetic engineering depends to a large extent on the ability of classical biotechnology to implement the achievements of the former in the field. To prove its worth in the ''real world'', it will be the task of environmental engineering biotechnology to render itself applicable in outdoor, large-scale, nonsterile reactors.

V. CONCLUSIONS

Environmental engineering practice will have to resort to more advanced and more efficient biological treatment techniques to treat organic wastes. Special emphasis should be given to the increasing problem of slow biodegradability of recalcitrant compounds, as well as trying to remove inorganic compounds such as selenium from agricultural drainage waters using biological methods.

Environmental engineering biotechnology should use the new advances in both genetic engineering and classical biotechnology and should therefore be recognized as an important branch of biotechnology.

Chapter 5

MODERN BIOLOGICAL METHODS: THE ROLE OF BIOTECHNOLOGY

Gary S. Sayler and James W. Blackburn

TABLE OF CONTENTS

I. INTRODUCTION

Over the past two decades, developments in recombinant DNA technology have promoted a virtual explosion of research and new knowledge in modern molecular biology. The rapid development of this field is the result of scientific breakthroughs allowing the controlled modification, introduction, and expression of foreign or native genes in a host organism. The recombinant DNA technology used to reach these achievements was the subject of serious scientific and public concern over the risks and ethics of such "genetic engineering". These concerns and their possible impediments to future research were expressed at numerous scientific forums, perhaps best exemplified by the Asilomar Conference,[1] and led to the eventual establishment of the recombinant DNA advisory committee (RAC) and the National Institutes of Health (NIH) guidelines for recombinant DNA research. The flexibility and evolution of the NIH guidelines, in response to an expanding knowledge base of the limitations and safety of recombinant DNA technology, have been largely responsible for the structured growth and transition from basic and applied research to the development and use of living organisms and their parts and processes to benefit mankind. Opportunities abound for the use of "tailor-made" microorganisms and their protein products in product-oriented areas such as health care, agriculture, commodity and specialty chemicals, and environmental protection.[2]

The direct use of microorganisms and their capabilities to solve environmental problems and for *in situ* agricultural and industrial applications can be defined operationally as environmental biotechnology. Applications include detoxification and/or destruction of pollutants and hazardous wastes, improvements in soil fertility and crop productivity, biological pest management, and restoration and renovation of perturbed ecosystems. Environmental biotechnology is differentiated from other areas of biotechnology in that successful process development must contend with the complexity of mixed (heterogeneous) populations and interactions occurring in ecosystems, as well as by the fact that the technology itself may be the subject of concern over environmental hazards. Consequently, there are major needs for both ecological and engineering research to enable utilization of modern applied molecular biology for successful process development and overall risk assessment.

The applications for molecular biology and recombinant DNA technology in the management of hazardous agricultural wastes and environmental decontamination fall into three general areas:

1. Isolation and microbial strain improvement for developing microorganisms with greater capacity for destruction of hazardous wastes and environmental contaminants
2. Field-site evaluation of microbial degradation processes contributing to overall contaminant fate predictions in a given system
3. Development, monitoring, and control of engineered processes for the biological destruction of hazardous waste and environmental contaminants

Each of these areas benefits directly from knowledge and research tools made available by molecular biology. However, it is the area of strain development or improvement which has received most of the popular, if not technical, attention. This attention has been directed toward research to develop genetically engineered microorganisms with new or improved biodegradative capabilities. While this is an important area of research with potentially significant contributions to developing biological treatment processes for difficult-to-degrade contaminants, molecular biology knowledge and recombinant DNA technology can also contribute to the development of nonengineered biodegradative microorganisms and processes. It can also be demonstrated that this same knowledge and technology will contribute, with even greater impact, to the successful understanding and utilization of microorganisms for hazardous waste control.

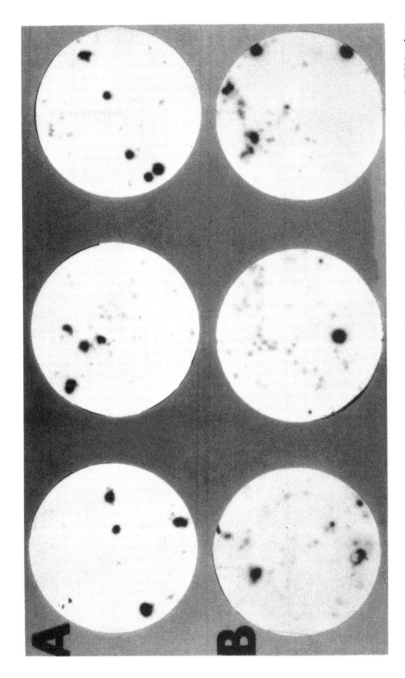

PLATE 1. ^{32}P autoradiographic detection of toluene and naphthalene catabolic genotypes in synthetic oil contaminated sediments by DNA-colony hybridization. (A), TOL plasmid positive hybrids; (B), NAH plasmid positive hybrids. Intensity of autoradiographic spot is related to both colony size and degree of homology with respective DNA probe. (From Sayler, G. S., Shields, M. S., Tedford, E. T., Breen, A., Hooper, S. W., Sirotkin, K. M., and David, J. W., *Appl. Environ. Microbiol.*, 49, 1295, 1985. With permission.)

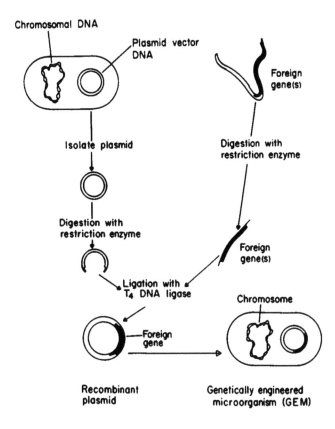

FIGURE 1. General approach for developing genetically engineered microorganisms for application in hazardous waste management, i.e., cloning of degradative foreign genes into a vector plasmid to create a recombinant DNA-containing microorganism. (From Jain, R. K., Burlage, R. S., and Sayler, G. S., *Crit. Rev. Biotechnol.*, 8, 33, 1988. With permission.)

II. DEVELOPMENT OF BIODEGRADATIVE STRAINS

A general example of genetically engineered microbial strains to increase biodegradative capacity is described in Figure 1. In this example, a bacterial cell containing a plasmid with known characteristics is cultivated, the chromosomal and plasmid DNA is isolated, and the plasmid DNA is separated in pure form. The plasmid is enzymatically cut with one or more specific restriction endonucleases to generate a linear molecule to which new biodegradative genes will be ligated. Biodegradative genes from a second bacterium are isolated in much the same manner from either the chromosome or a plasmid and, under *in vitro* conditions, are enzymatically joined to the original vector plasmid to reconstitute the circular recombinant DNA plasmid molecule. The new plasmid is then reintroduced, via transformation, to the original (or a secondary) bacterial host to yield a genetically engineered microorganism with new biodegradative capacity. This general scheme can be modified in a variety of ways, resulting in the introduction of regulatory or structural genes into either the bacterial chromosome or another plasmid using a variety of plasmid cloning vectors.

A more specific example of such recombinant plasmids is given by the construction of a hybrid plasmid containing genes from the toluene/xylene (TOL) catabolic plasmid, pWWO (Figure 2).[4] This hybrid plasmid (pPL403) contains the regulatory *xyl*S from the TOL plasmid, as well as the structural *xyl*D and *xyl*L genes coding for toluate oxygenase and dihydroxybenzoate dehydrogenase, respectively. This hybrid plasmid, maintained in *Esch-*

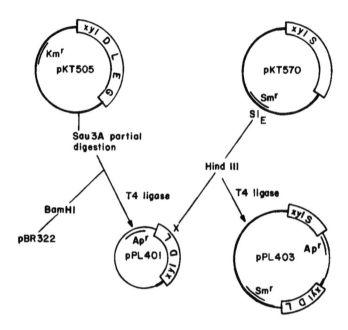

FIGURE 2.   Molecular construction of the hybrid plasmid pPL403 containing the TOL plasmid (pWWO) genes *xylS*, D, and L used to expand the range of chlorobenzoate metabolism in *Pseudomonas* B13 (WR1).[4] Thick lines indicate plasmid DNA from plasmid pKT231, thin lines represent DNA from plasmid pBR322, and boxes indicate TOL plasmid DNA from pWWO. Ap[r], Km[r], and Sm[r] indicate genes coding for resistance to the antibiotics ampicillin, kanamycin, and streptomycin, respectively.

erichia coli K12 strain ED 8654, was mobilized by a second plasmid, RP1, and conjugatively transferred into *Pseudomonas* strain B13 (WR1). The WR1 strain was normally able to degrade and grow on 3-chlorobenzoate, but not 4-chlorobenzoate (4-CBZ) or 3,5-dichlorobenzoate (3,5-DCBZ). However, WR1 with plasmid pPL403 could degrade and grow on all three substrates. It was demonstrated that the TOL *xylD* gene was required for 4-CBZ degradation, but both *xylD* and *xylL* were required for 3,5-DCBZ degradation.

Such experiments are typical of those designed to engineer microorganisms to expand the range of pollutants on which they are able to be grown. In such an experiment, at least four separate plasmids were required for the described work: plasmid vectors for gene cloning, pKT231[5] and pBR322;[6] a mobilization plasmid, RP1;[7] and the TOL catabolic plasmid, pWWO.[8] Successful genetic engineering thus requires available genetic tools and a thorough understanding of the molecular biology of the organisms and plasmids used.

Plasmids are the most common vehicle for genetic engineering. Ranging in size from several thousand to several hundred thousand base pairs, these extrachromosomal genetic elements are generally not required for normal replication or maintenance functions of the microbial cell and can be gained or lost by the host cell with a concomitant gain or loss of the plasmid-associated phenotype or genotype. Many of the larger plasmids, in addition to containing genes associated with their own replication, contain genes that encode the ability to transfer from one bacterial cell to a compatible bacterial cell. Sometimes this transfer range, or host range, is very broad (across generic boundaries), resulting in designation as a broad-host-range plasmid. Related plasmids that share similar replication origins in terms of molecular structure or have similar partitioning characteristics generally are not maintained simultaneously in the same cell and are considered incompatible and, thus, related in families of incompatibility groups. Much of the fundamental plasmid molecular biology has been related to enteric bacterial hosts. However, in recent years, extensive research on plasmids in other Gram-negative bacteria, in Gram-positive bacteria, and in fungi has been conducted.

An important finding of this research is that in a variety of terrestrial and aquatic bacterial species (most notably *Pseudomonas, Alcaligenes,* and *Acinetobacter*), biodegradation of recalcitrant synthetic and naturally occurring organic compounds is plasmid encoded. An extensive listing of naturally occurring catabolic plasmids is given in Table 1. All of these plasmids have been described since Chakrabarty's description of the salicylate catabolic plasmid[9] in 1972. Among the earlier catabolic plasmids described were the pJP series of plasmids involved in degradation of the herbicides 2,4-D and MCPA.[10,11] The fact that these herbicides are not long-lived in soil may be in part attributable to such naturally occurring plasmids; 2,4,5-T, a closely related herbicide, is persistent in many environments, and as yet no naturally occurring plasmid has been isolated for this contaminant.

The complexity of these catabolic plasmids, which are generally in the larger size range for plasmids, is described by the TOL plasmid[8,12] (Figure 3). This well-characterized plasmid, 117 kb in size, encodes the degradation of toluene, *m-* and *p*-toluate, and *m-* and *p*-xylene to tricarboxylic acid cycle intermediates. Catabolism of these substrates is regulated by two genes, *xyl*R and *xyl*S. The upper pathway, *xyl*CAB, encodes the genes for degradation of toluene and xylenes to aromatic carboxylic acids and is positively regulated by *xyl*R. These carboxylic acid products activate the *xyl*S gene, which is the positive regulatory gene of the lower pathway, *xyl*DLEGFJIH, allowing further degradation to tricarboxylic acid cycle intermediates. The *xyl*R gene is able to act in concert with *xyl*S to augment the lower pathway activity. In addition to these catabolic genes, the plasmid also encodes its own transfer/replication (tra/rep) functions (see Figure 3).

Given the relative youth of investigations on the occurrence of plasmids in nature and the molecular characterization of newly discovered plasmids, there is no doubt that further research in this area will greatly enhance the potential to develop organisms with increased activity and broadened specificity for degradation of environmental contaminants. Knowledge derived from this area of research has already been applied to develop methods for the selective enrichment of degradative bacterial strains with new catabolic potential. One approach, termed plasmid-assisted molecular breeding, has been used by Kellogg et al.[13] to select bacterial cultures newly capable of 2,4,5-T degradation. This method utilizes chemostat cultivation with intense selective pressure and organisms with known catabolic plasmids to enhance the apparent natural evolution and gene recruitment, resulting in new genetic combinations for increased degradative performance. Similar selection had previously shown the ability to obtain new isolates for haloaromatic metabolism.[14]

III. MOLECULAR DETECTION AND MONITORING

A relatively new approach for the detection and isolation of organisms with catabolic potential utilizes nucleic acid hybridization techniques (Figure 4). These techniques, sometimes termed gene probing, were developed for detection and screening of recombinant organisms in laboratory experiments and are now finding utility in medical diagnostics. This technology relies on the ability to isolate, purify, and label DNA of known catabolic origin to prepare a specific probe. The double helical structure of the probe DNA is destroyed by heating to create single-stranded probe nucleic acid. The probe is then added to either similarly treated target DNA from a bacterial colony or to an extract of DNA (from an environmental sample) which has been bound to a hybridization filter. Under the proper conditions, the probe and target DNA are allowed to reassociate to reform the double helical structure. Reassociation of the probe with complementary strands of the target nucleic acid results in a hybrid molecule that is readily detected by the probe-associated label (e.g., ^{32}P, ^{3}H, ^{35}S, antigen-antibody complexes, and enzyme-substrate reactions).[3] This hybridization can be controlled for highly specific detection of nucleic acids complementary to the probe. This technology has been successfully used to detect specific catabolic genotypes in envi-

TABLE 1
Naturally Occurring Catabolic Plasmids[39]

Primary substrate	Plasmid	Size (kb)	Incompatibility group	Original host	Transmissibility[a]
6-Aminohexanoate cyclic dimer	pOAD2			*Flavobacterium* spp. K172	
Acetate (2-chloro-)	pUO1	100	ND	*Moraxella* spp. B	+
Acetate (2-fluoro-)	pUO1	100	ND	*Moraxella* spp. B	+
Benzene (1,2,4-trimethyl-)	pDK1	180	ND	*Pseudomonas putida* 11S1 (PpC1)	+
Benzoate (3,5-dichloro-)	pAC29	72	ND	Derivative of pAC27	ND
Benzoate (3-chloro-)	pAC25	117	ND	*P. putida* AC25	+
Benzoate (3-chloro-)	pAC25			*P. putida*	
Benzoate (3-chloro-)	pAC27	110	ND	Derivative of pAC25	+
Benzoate (3-chloro-)	pJP3	78.1	P-1	*Alcaligenes paradoxus*	+
Benzoate (3-chloro-)	pJP4	78.1	P-1	*Alcaligenes eutrophus*	+
Benzoate (3-chloro-)	pJP5	78.1	P-1	*A. eutrophus*	+
Benzoate (3-chloro-)	pJP7	78.1	P-1	*A. eutrophus*	+
Benzoate (3-chloro-)	pWRI			*Pseudomonas* sp. B13	
Benzoate (4-chloro-)	pAC27	110	ND	Derivative of pAC25	+
Benzoate (4-chloro-)	pSS50	53.2	Not I,P,Q,W	*Alcaligenes* spp. A2, A2D, A5	+
Benzoate (4-chloro-)	pSS50	53.2	Not I,P,Q,W	*Acinetobacter* spp. A8, AX2	+
Biphenyl	pBS241	ND	ND	*P. putida* BS893	ND
Biphenyl (2,2'-dihydroxy-)	pBS241	ND	ND	*P. putida* BS893	ND
Biphenyl (3,3'-dimethyl-)	pBS241	ND	ND	*P. putida* BS893	ND
Biphenyl (4,4'-dichloro-)	pBS241	ND	ND	*P. putida* BS893	ND
Biphenyl (4,4'-dimethyl-)	pBS241	ND	ND	*P. putida* BS893	ND
Biphenyl (4-chloro-)	pKF1			*Acinetobacter* sp. P6	
Biphenyl (4-chloro-)	pSS50	53.2	Not I,P,Q,W	*Alcaligenes* spp. A2, A2D, A5	+
Biphenyl (4-chloro-)	pSS50	53.2	Not I,P,Q,W	*Acinetobacter* spp. A8, AX2	+
Camphor	CAM			*P. putida*	
Cresol (*p*-)	pND50			*P. putida*	
Cresol (*p*-)	pRA1000	85	P-9	*Alcaligenes eutrophus* 345	+
Diphenylmethane	pBS241	ND	ND	*P. putida* BS893	ND
Methylbenzyl alcohol (*m*-)	pWWO-152	116	P-1 (RB4)	*P. putida* PaW152	+
Naphthalene	NAH7	72	P-9	*P. putida* PpG7	+
Naphthalene	pBS2			*P. putida* BS240	

TABLE 1 (continued)
Naturally Occurring Catabolic Plasmids[39]

Primary substrate	Plasmid	Size (kb)	Incompatibility group	Original host	Transmissibility[a]
Naphthalene	pBS3			P. putida BS245	
Naphthalene	pWW60-1			P. putida	+
Nicotinate	NIC-T			P. convexa	
Nicotine	NIC-T			P. convexa	
Octane (-n,C6-C10 alkane)	OCT			P. putida	
Parathion	pCSI	66	ND	P. diminuta	ND
Phenoxyacetate (2,4-dichloro-)	pJP12	87		Alcaligenes paradoxus	ND
Phenoxyacetate (2,4-dichloro-)	pJP2	55.3	P-1	A. paradoxus	
Phenoxyacetate (2,4-dichloro-)	pJP3	78.1		A. paradoxus	+
Phenoxyacetate (2,4-dichloro-)	pJP4	78.1	P-1	A. eutrophus	+
Phenoxyacetate (2,4-dichloro-)	pJP5	78.1	P-1	A. eutrophus	+
Phenoxyacetate (2,4-dichloro-)	pJP7	78.1	P-1	A. eutrophus	+
Phenoxyacetate (2,4-dichloro-)	pJP9	55.3	P-1	A. eutrophus	ND
Phenoxyacetate (4-chloro-2-methyl-)	pJP1	87		A. paradoxus	
Phenoxyacetate (4-chloro-2-methyl-)	pJP2	55.3	P-1	A. paradoxus	ND
Phenoxyacetate (4-chloro-2-methyl-)	pJP3	78.1		A. paradoxus	
Phenoxyacetate (4-chloro-2-methyl-)	pJP4	78.1	P-1	A. eutrophus	+
Phenoxyacetate (4-chloro-2-methyl-)	pJP5	78.1	P-1	A. eutrophus	+
Phenoxyacetate (4-chloro-2-methyl-)	pJP9	55.3	P-1	A. eutrophus	ND
Phenylacetate	pWW17			P. putida	
Proprionate (2-chloro-)	pUU204			Pseudomonas sp. EA	
Salicylate	SAL			P. putida	+
Salicylate	pAC10::SAL	84	P-1 (RP4)	P. putida AC843	+
Salicylate	pKF439	138	P-9 (SAL)	P. putida KF439	+
Salicylate	pMWD1			P. putida	
Toluate (m-)	R702::TOL		(R702)	P. putida 839	+
Toluate (m-)	RP4::TOL	114	P-1 (RP4)	P. aeruginosa PAO	+
Toluate (m-)	TOL	28	P-9	P. putida AC10	−
Toluate (m-)	TOL-TOL	111	P-9	P. putida AC802	+
Toluate (m-)	TOL K	162	(K)	P. putida AC797	+
Toluate (m-)	pKF439	138	P-9 (SAL)	P. putida KF439	+
Toluate (m-)	pND2	ND	(R-91)	P. aeruginosa PAO3	+
Toluate (m-)	pND3	ND	(R-91)	P. aeruginosa PAO3	+
Toluate (m-)	pRA1000	85	P-9	Alcaligenes eutrophus 345 (ATCC 17707)	+
Toluate (m-)	pTKO1	150	ND	P. putida PPK1	−
Toluate (m-)	pTN001	117	P-9	P. putida mt-2	+

TABLE 1 (continued)
Naturally Occurring Catabolic Plasmids[39]

Primary substrate	Plasmid	Size (kb)	Incompatibility group	Original host	Transmissibility[a]
Toluate (*m*-)	pTN1	114	P-1 (RP4)	*P. putida* TH5004	+
Toluate (*m*-)	pTN2	111	P-1 (RP4)	Derivative of pTN1	+
Toluate (*m*-)	pTN8	114	P-1 (RP4)	Derivative of pTN1	+
Toluate (*m*-)	pTN81	109	P-1 (RP4)	Derivative of pTN1	+
Toluate (*m*-)	pTN9	91	P-1 (RP4)	Derivative of pTN1	+
Toluate (*m*-)	pWWO-152	116	P-1 (RP4)	*P. putida* PaW152 (ATCC 17707)	−
Toluate (*m*-)	pWW53	107	Not P-1,9	*P. putida* NT53	+
Toluate (*p*-)	pND2	ND	(R-91)	*P. aeruginosa* PAO3	+
Toluate (*p*-)	pND3	ND	(R-91)	*P. aeruginosa* PAO3	+
Toluate (*p*-)	pRA1000	85	P-9	*Alcaligenes eutrophus* 345 (ATCC 17707)	+
Toluene	TOL (pWWO)	117	P-9	*P. putida* mt-2	+
Toluene	XYL-K			*Pseudomonas* pxy	+
Toluene	pDK1	180	ND	*P. putida* HS1 (PpC1)	ND
Toluene	pKJ1	150	Not P-1,2,3,9	*Pseudomonas* sp. TAB	+
Toluene	pND2	ND	(R-91)	*P. aeruginosa* PAO3	+
Toluene	pND3	ND	(R-91)	*P. aeruginosa* PAO3	+
Toluene	pTN001	117	P-9	*P. putida* mt-2	+
Toluene	pTN2	111	P-1 (RP4)	Derivative of pTN1	+
Toluene	pTN8	114	P-1 (RP4)	Derivative of pTN1	+
Toluene	pTN81	109	P-1 (RP4)	Derivative of pTN1	+
Toluene (3-ethyl-)	pDK1	130	ND	*P. putida* HS1 (PpC1)	+
Xylene	XYL-K			*Pseudomonas* pxy	+
Xylene (*m*-)	TOL (pWWO)	117	P-9	*P. putida* mt-2	+
Xylene (*m*-)	pDK1	180	ND	*P. putida* HS1 (PpC1)	+
Xylene (*m*-)	pKF439	138	P-9 (SAL)	*P. putida* KJ439	ND
Xylene (*m*-)	pKJI	150	Not P-1,2,3,9	*Pseudomonas* sp. TAB	+
Xylene (*m*-)	pND2	ND	(R-91)	*P. aeruginosa* PAO3	+
Xylene (*m*-)	pND3	ND	(R-91)	*P. aeruginosa* PAO3	+

TABLE 1 (continued)
Naturally Occurring Catabolic Plasmids[39]

Primary substrate	Plasmid	Size (kb)	Incompatibility group	Original host	Transmissibility[a]
Xylene (*m*-)	pTN001	117	P-9	*P. putida* mt-2	+
Xylene (*m*-)	pTN2	111	P-1 (RP4)	Derivative of pTN1	+
Xylene (*m*-)	pTN8	114	P-1 (RP4)	Derivative of pTN1	+
Xylene (*m*-)	pTN81	109	P-1 (RP4)	Derivative of pTN1	−
Xylene (*m*-)	pWW53	107	Not P-1,9	*P. putida* MT53	+
Xylene (*p*-)	TOL (pWWO)	117	P-9	*P. putida* mt-2	+
Xylene (*p*-)	pDK1	180	ND	*P. purida* HS1 (PpC1)	ND
Xylene (*p*-)	pKJ1	150	Not P-1,2,3,9	*Pseudomonas* sp. TAB	
Xylene (*p*-)	pND2	ND	(R-91)	*P. aeruginosa* PAO3	
Xylene (*p*-)	pND3	ND	(R-91)	*P. aeruginosa* PAO3	
Xylenol (3,5-)	NIC-T			*P. convexa*	

Note: ND = not determined; in cases where this was not explicitly stated, a blank occurs.

[a] Parenthetical notations in incompatibility column indicate the resistance plasmid which was used to create the fusion.

ronmental samples[15] (Plate 1)* and to isolate new bacterial strains with broader ranges of degradable substrates.[16] Some proven applications of this technology are given in Table 2.

One example that demonstrates the utility of this gene-probe detection and monitoring technology is the use of gene probes to study polychlorinated biphenyl (PCB) degradation processes in reservoir sediments. A PCB catabolic plasmid (pSS50) isolated from bacteria present in PCB-contaminated reservoir sediment has been shown to encode a pathway for the complete oxidation of 4-chlorobiphenyl to CO_2.[17] Using this plasmid as a plasmid gene probe, colony hybridization experiments have demonstrated that 0.3% of the total aerobic bacteria in contaminated sediment are presumptively able to degrade 4-chlorobiphenyl.[16] The turnover time for 4-chlorobiphenyl in these sediments was previously shown to be 17 d ($k = 0.06$ d^{-1}), with the primary end product being CO_2.[22] When examining the individual cultures probing positive with the pSS50 plasmid, it has been demonstrated that all are capable of 4-chlorobiphenyl metabolism.[16] Furthermore, as indicated by Table 3, many of the cultures have broad growth ranges on different PCB substrates, including tetrachlorobiphenyl. It is anticipated that the ability to monitor specific degradative genotypes in complex environmental matrices such as soils will lead to a much greater predictive capability of overall fate processes for contaminants such as pesticides. A variety of developing methods for detection of engineered organisms in the environment should prove suitable for such purposes (Table 4).[3] The availability of such technology will allow prediction of biodegra-

* Plate 1 follows page 54.

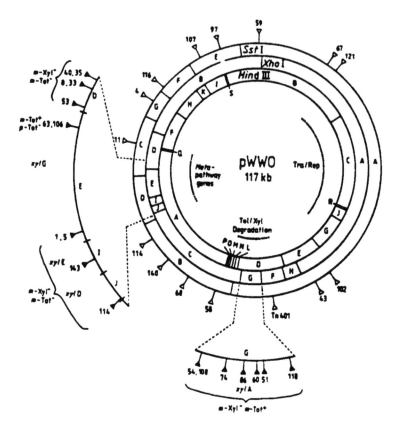

FIGURE 3. Functional map of the toluene/xylene catabolic plasmid pWWO. (The regulatory genes *xyl*R and S are adjacent and distal to the *xyl*D, L, E, F, and G genes, as indicated; *xyl*L is internal to *xyl*D and E.) (From Franklin, F. C. H., Lehrbach, P. R., Lurry, R., Rueckert, B., Bagdasarian, M., and Timmis, K. N., *Proc. Natl. Acad. Sci. U.S.A.*, 78, 7458, 1981. With permission.)

dation rates to optimize pesticide application rates as well as identification of problem soils for which alternative pest management scenarios should be encouraged.

IV. TREATMENT PROCESS MONITORING AND CONTROL

As has been indicated, the potential has been demonstrated to develop improved strains for degradation of wastes and contaminants and to monitor the population dynamics of such organisms in the environment. The potential also exists to design treatment processes in which the population dynamics of specific degradative genotypes can be optimized to achieve predictable and efficient destruction of specific wastes and contaminants. The optimization of such dynamic systems, which can include numerous degradative and nondegradative genotypes, requires the ability to physically, chemically, or biologically control the system. This control then necessitates a monitoring capability for the degradative organisms of interest. A recent example of achieving genotype-level monitoring in a hazardous waste treatment process for naphthalene removal is available.[19] Using naphthalene-specific gene probes derived from the naphthalene-degradative plasmid NAH7 in *Pseudomonas putida*, the population dynamics of naphthalene-degradative genotypes were measured and correlated with overall system performance and naphthalene treatment fate. As indicated by Figure 5,

DNA PROBE FILTER HYBRIDIZATION

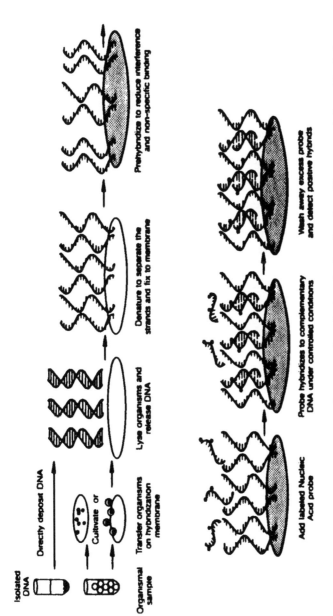

FIGURE 4. Protocol for DNA filter hybridization gene probe technology for detection of specific sequences in bacterial colonies or isolated DNA.[3]

TABLE 2
Applications of DNA Probe Technology in Biodegradation Research

Study	Probes	Results	Ref.
Adaptation to synthetic oils in sediments (PAH)	TOL plasmid, NAH plasmid	Correlation of rates of PAH mineralization with genotype frequency over time	15
Maintenance of introduced species in aquifer material	TOL plasmid, TOL *xylR* gene, cryptic genes	Long-term stable maintenance of catabolic genotype	18
Rapid isolation of catabolic genotypes	pSS50 plasmid (PCB biodegradation)	One week isolation of confirmed genotype from sediment	16
Activated sludge biodegradation of naphthalene	NAH plasmid, pDTG 113 plasmid (NAH clone)	Gene frequency inverse correlation with stripping; possible threshold	19
Biodegradation of PCB in soil by introduced genotypes	pSS50 plasmid	Correlation of mineralization with genotype frequency	38
Structure of biodegradative communities	*Nif* genes, NAH genes, pSS50 plasmid	Direct detection of specific genes in sludge and sediment extracts	20, 21

TABLE 3
Growth of Reservoir Sediment Isolates, Obtained by
pSS50 Colony Hybridization, on Chlorobiphenyls

Strain	Growth substrate[a]			
	4-CB	4,4'-CB	2,4,5-CB	2,4,2',4'-CB
LPS9	+ + +	+ + +	+ +	+
LPS9AB	+	+	+	+
LPS9C	+ + +[b]	+ + +[b]	+	+
A5 (pSS50)	+ + +[b]	+ + +[b]	+	+
A5 (pSS50⁻)	−	−	−	−

Note: Strains A5 (pSS50) and A5 (pSS50⁻) are positive and negative controls, respectively. Substrates: 4-CB = 4-chlorobiphenyl, 4,4'-CB = 4,4-dichlorobiphenyl, 2,4,5-CB = 2,4,5-trichlorobiphenyl and 2,4,2',4'-CB = 2,4,2',4'-tetrachlorobiphenyl.

[a]   Incubation for 10 d at 100 mg/l of chlorinated biphenyl substrate, as indicated.
[b]   Cultures turn bright yellow, suggesting the formation of high concentrations of the ring fission product.

the use of specific naphthalene degradation gene probes provided superior sensitivity and resolution of fluctuations in population densities of naphthalene-degradative genotypes in a completely mixed activated-sludge treatment process (Figure 5C). Increases in the critical naphthalene-degradative genotype were directly related to the mineralization fate and inversely related to the stripping fate of naphthalene in the treatment process (Figures 5A and B). The results of these studies may indicate a threshold in degradative cell number that, when achieved, drives the fate of naphthalene to biological mineralization at the expense of stripping.[19] The results of such studies can provide direction to determine system variables that can be adjusted to optimize naphthalene-degradative cell densities at or above this threshold in order to achieve maximum biological mineralization.

A variety of open and closed treatment systems for the biological destruction of residues in soil have been described;[23] many of these are amenable to molecular monitoring of critical

TABLE 4
Developing Methods for Monitoring Genetically
Engineered Microorganisms in the Environment with
Potential Application for Quantifying Degradative
Populations[3]

Developing Methods

Immunological techniques
Enzyme-linked immunosorbent assay (ELISA)
Radioactive markers
Fluorescent markers
Plasmid epidemiology and restriction profile
   Use of plasmid epidemiology
   Use of restriction profile (RFLP)
Selectable genotypic markers
Nucleic acid sequence analysis
   DNA sequence analysis
   Ribosomal RNA sequence analysis
Nucleic acid hybridization techniques
   DNA:DNA colony hybridization
   Southern blot hybridization
   Nucleic acid hybridization with DNA extracts
   DNA:RNA hybridization
   Biotinylated probes
New selective enrichments
Protein and enzyme analysis
   Isozymes (multilocus enzyme electrophoresis, MLEE)
   Protein gels (SDS polyacrylamide gel electrophoresis, SDS/PAGE)
Genetically engineered markers
   Metabolic changes
   Surface marker changes
   Susceptibility changes

gene frequencies. Soil composting systems may provide desired levels of control to permit optimization of biological degradation processes. Research in this area could also lead to defining system variables that can be manipulated in field systems to greatly enhance naturally occurring biodegradation or to establish engineered biodegradative processes. This potential capacity of microorganisms to degrade a variety of xenobiotic and recalcitrant wastes has been described in numerous reviews,[24-27] but exploitation of this vast degradation potential has yet to be realized.

V. SUMMARY AND RESEARCH NEEDS IN THE APPLICATION OF BIOTECHNOLOGY TO WASTE MANAGEMENT

The role of microorganisms in the biodegradation and biotransformation of natural organic and inorganic matter, as well as synthetic environmental contaminants and hazardous wastes, has long been recognized by environmental scientists and engineers. Historically, the biodegradative capacities of microorganisms have been taken advantage of in engineered systems to reduce the COD and BOD of agricultural, domestic, and industrial wastes in order to permit environmental discharge. Such applications have proven efficient, often achieving a greater than 90% reduction in the level of bulk parameters like COD. In a similar fashion, the biodegradative capabilities of microorganisms in nature have been viewed as a "self-purification" mechanism when environmental contamination does occur. However, with increased awareness of the toxicological significance of the recalcitrant organic compounds comprising a part of the 10% COD that may not be removed by conventional

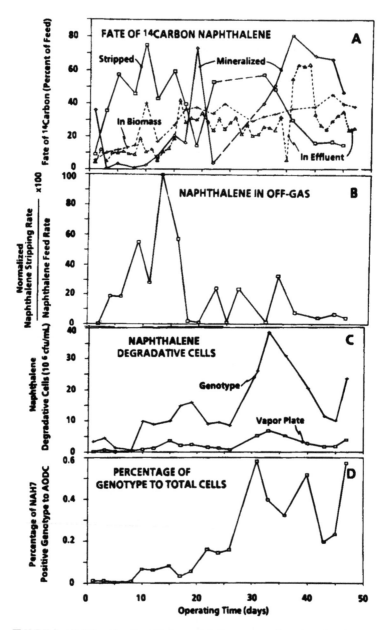

FIGURE 5. Relationship of naphthalene catabolic genotypes in activated sludge to naphthalene stripping and mineralization fate in the wastewater treatment process. (From Blackburn, J. W., Jain, R. K., and Sayler, G. S., *Environ. Sci. Technol.*, 29, 884, 1987. With permission.)

biological treatment and the fact that persistence and accumulation of contaminants in the environment is well documented, it has become necessary to reevaluate both engineering designs to achieve optimum system performance and the capacity of microbial populations responsible for the biochemical conversion of apparent recalcitrant compounds in wastewater and the environment. Changes in system design and operation, such as equilibration to reduce shock loading, O_2 injection, fixed film and rotary biological contactor (RBC) processes, and the sequencing batch reactor (SBR), have all contributed to increased levels of performance for some recalcitrant wastes. In a parallel fashion, manipulation of contaminated environments either nutritionally or hydrodynamically (such as in groundwater) to promote enhanced biodegradation has met with some measure of success. The success or lack of success of these improvements has been largely without specific knowledge of the microbial populations responsible for the observed biodegradation. Often there is a failure to appreciate or acknowledge that the biochemical capacity for degradation of specific recalcitrant organic molecules is not uniformly or ubiquitously distributed among microbial populations or, for that matter, among individuals of the same species.

As a result, limitations in the application of conventional biological treatment exist that affect the future success of biological hazardous waste management and environmental decontamination. These limitations are summarized as follows:

1. Contemporary applications are limited to relatively easy-to-degrade compounds — the availability of organisms or systems for xenobiotic compounds is low.
2. Contemporary applications have limited predictability across different waste types or environments — they are relatively uncharacterized systems with few specific predictive parameters.
3. Contemporary applications are limited in specific process control — they are relatively uncharacterized systems with few specific control parameters.
4. Performance standards are largely unvalidated by scientific methods — alternative fate processes and interactions are frequently not quantified.

The economic attractiveness and potential efficiency of biodegradation of hazardous wastes are recognized. But, even in the nontechnical popular literature, limitations are recognized that affect the applications for native or engineered microorganisms in hazardous waste control.[28,29] These are summarized as follows:

1. The relatively limited availability of degradative strains with proven biochemical capacity
2. The need for specific biochemical and genetic knowledge of biodegradation for strain improvement or modification
3. The need for specific ecological knowledge to permit strains to persist and function in ecologically complex and sometimes toxic environments

Some of these issues — the mass of circumstantial evidence that is used to support assumed biodegradation in contaminated environments, the limitation of existing biodegradation data bases, and limitations of current microbiological techniques in environmental measurement — have been discussed in recent reviews of biodegradation in groundwater and activated sludge.[30,31] It has also been noted that neither federal nor commercial funding has been directed toward long-term solutions to these needs and limitations.[29] Obstacles to implementing biotechnology in hazardous waste reduction are likely to be ''institutional and behavioral'' rather than technical, as suggested by a recent Office of Technology Assessment (OTA) report.[37]

Over the past 2 decades, an increased understanding of the molecular basis for biodegradation of select groups of environmental contaminants has emerged. In conjunction, developments in modern molecular biology and recombinant technology have occurred that permit a much faster pace in elucidating the mechanisms of biodegradation of other contaminants. These combined developments make possible the planning of research to genetically modify microorganisms to expand the range of substrates degraded, as well as to increase the relative rate at which degradation occurs. This specific topic and the applications of genetics to the degradation of hazardous wastes and environmental contaminants have been the subject of several workshops and symposia.[32-35] A distillation of the collective discussions can be summarized as two major points:

1. The potential exists for degradative strain expansion and improvement using advanced molecular techniques.
2. This potential is not being realized in terms of basic science and engineering research or reduction to practical application.

At a more fundamental level, new molecular approaches are also available that permit specific analysis of biodegradative genotypes in heterogeneous microbial populations in complex environmental matrices[15,19] (see also Table 4). With the availability of sensitive quantitative measures of specific degradative populations, it is possible to develop research strategies to determine specific control parameters that influence microbial population densities and activity and ultimately define the limits of obtainable process optimization.[28,31] Potential advances that can be made in process optimization are independent of specific efforts to genetically engineer microbial strains and can be applied to virtually all naturally occurring microbial populations in natural or engineered systems.

The governing hypothesis of research needs in this area is that the contributions of environmental biotechnology to hazardous waste management and biodegradation of environmental contamination have not been realized due to:

1. The need for more fundamental information on the occurrence, biochemistry, and genetics of biodegradative microorganisms for specific environmental contaminants
2. The need for developing a cross-disciplinary research infrastructure that integrates the molecular biology, microbial ecology, and engineering science required for biodegradative process development and optimization

It is the objective of environmental biotechnology to bring together cross-disciplinary experts in biodegradation research and engineering to overcome these current research limitations for implementation of the technology and to delineate a set of basic science and engineering research imperatives required to exploit the technology. A second objective is to provide advanced training for scientists and engineers in environmental biotechnology for implementing the technology for control of hazardous wastes and environmental contaminants.

These objectives are at the interface of science and engineering disciplines and require cross-disciplinary collaborative efforts and support by institutions and funding organizations. Such integrated collaborative efforts are generally rewarding, but, as discussed by the National Research Council, may be hampered by institutional opinions and barriers:

Interdisciplinary research between physical-engineering sciences and biological-clinical sciences has produced important new knowledge and medical advances, including research and diagnostic instruments such as CAT scans and permanent implant devices such as cardiac pacemakers and joint replacements. Applying the knowledge of how biological systems are designed promises to result in more refined and powerful techniques to further basic understanding and to treat living organisms.

But there are some major impediments to interdisciplinary collaborations. Differences in conceptual approach may create language and communication barriers. Lack of career incentives and rewards due to institutional and organizational constraints also may interfere. And formal training and orientation are not the same among practitioners of different disciplines. Because of these differences, scientists may be unaware of how and when to capitalize on the benefits to be gained by collaborative research.[36]

Basic science and engineering research is needed to exploit environmental biotechnology for biological treatment of hazardous wastes and environmental contaminants. Cross-disciplinary research representing microbial biochemistry, genetics, ecology, environmental microbiology, and chemical and environmental bioprocess engineering is imperative to advance biological waste treatment and to insure that the field is successfully integrated in the overall strategy of process development for management and control of hazardous wastes and environmental contaminants. The goal of environmental biotechnology should be to provide specific current and future direction for integrating modern biotechnology in basic environmental science and engineering research in the control of hazardous wastes, be they agricultural, domestic, or industrial.

REFERENCES

1. International Conference on Recombinant DNA Molecules, Asilomar Conference Center, Pacific Grove, CA, February 1975.
2. **Office of Technology Assessment,** Commercial Biotechnology: An International Analysis, OTA-BA-218, Office of Technology Assessment, Washington, D.C., January 1984.
3. **Jain, R. K., Burlage, R. S., and Sayler, G. S.,** Methods for detecting recombinant DNA in the environment, *Crit. Rev. Biotechnol.,* 8, 33, 1988.
4. **Lehrbach, P. R., Zeyer, J., Reineke, W., Knackmuss, H. J., and Timmis, K. N.,** Enzyme recruitment *in vitro:* use of cloned genes to extend the range of haloaromatics degraded by *Pseudomonas* sp. strain B13, *J. Bacteriol.,* 158, 1025, 1984.
5. **Bagdasarian, M. and Timmis, K. N.,** Host:vector systems for gene cloning in *Pseudomonas, Curr. Top. Microbiol. Immunol.,* 96, 47, 1982.
6. **Bolivar, F., Rodriguez, R. L., Green, P. J., Betlach, H. C., Heynecker, H. L., Boyer, H. W., Crosa, J. H., and Falkow, S.,** Construction and characterization of new cloning vehicles. II. A multipurpose cloning system, *Gene,* 2, 95, 1977.
7. **Bagdasarian, M., Lurry, R., Rueckert, B., Franklin, F. C. H., Bagdasarian, M. M., Frey, J., and Timmis, K. N.,** Specific purpose cloning vectors. II. Broad host range, high copy number RSF1010-derived vectors, and a host-vector system for gene cloning in *Pseudomonas, Gene,* 16, 737, 1981.
8. **Williams, P. A. and Murray, K.,** Metabolism of benzoate and methylbenzoates by *Pseudomonas putida* (arvilla) mt-2: evidence for the existence of a TOL plasmid, *J. Bacteriol.,* 120, 416, 1974.
9. **Chakrabarty, A. M.,** Genetic basis of the biodegradation of salicylate in *Pseudomonas, J. Bacteriol.,* 112, 815, 1972.
10. **Pemberton, J. M. and Fisher, P. R.,** 2,4-D plasmids and persistence, *Nature,* 268, 732, 1977.
11. **Don, R. H. and Pemberton, J. M.,** Properties of six pesticide degradation plasmids isolated from *Alcaligenes paradoxus* and *Alcaligenes eutrophus, J. Bacteriol.,* 145, 681, 1981.
12. **Franklin, F. C. H., Lehrbach, P. R., Lurry, R., Rueckert, B., Bagdasarian, M., and Timmis, K. N.,** Molecular and functional analysis of the TOL plasmid pWWO from *Pseudomonas putida* and cloning of genes for the entire regulated aromatic ring cleavage pathway, *Proc. Natl. Acad. Sci. U.S.A.,* 78, 7458, 1981.
13. **Kellogg, S. T., Chatterjee, D. K., and Chakrabarty, A. M.,** Plasmid-assisted molecular breeding: new technique for enhanced biodegradation of persistent toxic chemicals, *Science,* 214, 1133, 1981.
14. **Reineke, W. and Knackmuss, H. J.,** Construction of haloaromatic utilizing bacteria, *Nature,* 277, 285, 1979.
15. **Sayler, G. S., Shields, M. S., Tedford, E. T., Breen, A., Hooper, S. W., Sirotkin, K. M., and David, J. W.,** Application of DNA:DNA colony hybridization to the detection of catabolic genotypes in environmental samples, *Appl. Environ. Microbiol.,* 49, 1295, 1985.
16. **Pettigrew, C. A. and Sayler, G. S.,** The use of DNA:DNA colony hybridization in the rapid isolation of 4-chlorobiphenyl degradative bacterial phenotypes, *J. Microbiol. Methods,* 5, 205, 1986.
17. **Shields, M. S., Hooper, S. W., and Sayler, G. S.,** Plasmid mediated mineralization of 4-chlorobiphenyl, *J. Bacteriol.,* 163, 882, 1985.
18. **Jain, R. K., Sayler, G. S., Wilson, J. T., Houston, L., and Pacia, D.,** Maintenance and stability of introduced genotypes in ground water aquifer material, *Appl. Environ. Microbiol.,* 53, 996, 1987.
19. **Blackburn, J. W., Jain, R. K., and Sayler, G. S.,** The molecular microbial ecology of a naphthalene degrading genotype in activated sludge, *Environ. Sci. Technol.,* 29, 884, 1987.
20. **Sayler, G. S., Jain, R. K., Houston, L., Ogram, A., Pettigrew, C. A., Blackburn, J. W., and Riggsby, W. S.,** Applications for DNA probes in biodegradation research, in *Perspectives in Microbial Ecology,* Proc. 4th Int. Conf. Microbial Ecology, Megusar, F. and Ganter, M., Eds., Slovene Society for Microbiology, Ljubljana, Yugoslavia, 1988, 499.
21. **Ogram, A., Sayler, G. S., and Barkay, T.,** The purification and extraction of microbial DNA from sediment, *J. Microbiol. Methods,* 7, 57, 1988.
22. **Kong, H. L. and Sayler, G. S.,** Degradation and total mineralization of monohalogenated biphenyls in natural sediment and mixed bacterial culture, *Appl. Environ. Microbiol.,* 46, 666, 1983.
23. **Hanstveit, A. O., van Gemert, W. J. Th., Janssen, D. B., Rulkens, W. H., and van Veen, H. J.,** *Literature Study on the Feasibility of Microbiological Decontamination of Polluted Soils,* Netherlands Organization for Applied Scientific Research, TNO, The Hague, 1984, 44.
24. **Rochkind, M. L., Sayler, G. S., and Blackburn, J. W.,** *Microbiological Decomposition of Chlorinated Aromatic Compounds,* Marcel Dekker, New York, 1987.

25. **Leisinger, T., Cook, A. M., Hutter, R., and Neusch, J.,** *Microbial Degradation of Xenobiotics and Recalcitrant Compounds,* Academic Press, London, 1981.
26. **Chakrabarty, A. M.,** *Biodegradation and Detoxification of Environmental Pollutants,* CRC Press, Boca Raton, FL, 1982.
27. **Gibson, D. T.,** *Microbial Degradation of Aromatic Compounds,* Marcel Dekker, New York, 1986.
28. **Jain, R. K. and Sayler, G. S.,** Problems and potential for *in situ* treatment for environmental pollutants by engineered microorganisms, *Microbiol. Sci.,* 4, 59, 1987.
29. **Nicholas, R. B.,** Biotechnology in hazardous waste disposal: an unfulfilled promise, *Am. Soc. Microbiol. News,* 53, 138, 1987.
30. **Blackburn, J. W., Troxler, W. L., Truong, K. N., Zink, R. P., Meckstroth, S. C., Blorance, J. R., Groen, A., Sayler, G. S., Beck, R. W., Minear, R. A., Yagi, O., and Breen, A.,** Organic Chemical Fate Prediction in Activated Sludge Treatment Processes, EPA/600/S285/102, National Technical Information Service, Springfield, VA, 1985.
31. **Healy, J. B., Jr. and Daughton, C. G.,** Issues Relevant to Biodegradation of Energy-Related Compounds in Ground Water: A Literature Review, Contract #DE-ACO3-76SF0098, U.S. Department of Energy, Bartlesville, OK, 1986.
32. Biotechnology and Pollution Control, Proc. U.S. Environmental Protection Agency Workshop, Rep. No. 68-02-3952, U.S. EPA, Washington, D.C., 1986.
33. **Johnston, J. B. and Robinson, S. G.,** *Genetic Engineering and the Development of New Pollution Control Technologies,* Noyes Press, Park Ridge, NJ, 1984.
34. **Kulpa, C. F., Jr., Irvine, R. L., and Sojka, S. A.,** Proc. Symp. Impact of Applied Genetics in Pollution Control, University of Notre Dame, Notre Dame, IN, May 24 to 26, 1982.
35. **Parkin, G. F., Pipes, W. O., and Koerner, R. M.,** Research Needs Workshop: Hazardous Waste Treatment and Disposal. Proceedings of the Workshop, Drexel University, Philadelphia, June 9 to 10, 1986.
36. **U.S. National Research Council,** Interdisciplinary research collaboration, *Res. News,* 37, 27, 1987.
37. **Office of Technology and Assessment,** Serious Reduction of Hazardous Waste, OTA/GPO 052-003-01048-8, U.S. Government Printing Office, Washington, D.C., 1986.
38. **Blackburn, J. W. and Sayler, G. S.,** unpublished data.
39. **Hooper, S. W.,** Characterization of pSS50, the 4-Chlorobiphenyl Mineralization Plasmid, Ph.D. thesis, University of Tennessee, Knoxville, 1987.

Chapter 6

BACTERIAL DECONTAMINATION OF AGRICULTURAL WASTEWATERS

Ronald L. Crawford and Kirk T. O'Reilly

TABLE OF CONTENTS

I. INTRODUCTION

"Agricultural wastewaters" is a generic phrase that means different things to different people. The term is as diverse as the discipline of agriculture itself. To some, the term engenders visions of muddy streams choked with topsoil washed by spring rains from the slopes of carelessly cultivated fields. Others think of the dangers of nitrate leached into aquifer-based drinking water supplies from repeatedly fertilized farmlands. Perhaps a majority of individuals, however, associate agricultural wastewaters with pollution by anthropogenic chemicals, in recent times more often called "xenobiotic" compounds. As pointed out by McCormick,[1] the U.S. Environmental Protection Agency lists more than 60,000 chemicals marketed in the U.S., with more than 1,000 new compounds added to the list yearly. Thousands of these compounds are toxic and pose threats to public health. Many of these chemicals are produced for use in agriculture. Due to the very nature of agricultural enterprises, when toxic chemicals are employed they end up widely distributed. Unfortunately, water and/or soil often are primary repositories for toxic compounds employed in the businesses of producing food and fiber, and the contamination of soil and water resources is often vast, as in the case of chemical pollution of streams and aquifers.

So then, what can be done to remedy pollution of water and soil by hazardous chemicals when the contamination comes from nonpoint sources and encompasses vast volumes which are often in inaccessible places? We would like to suggest that the utilization of pollutant-degrading bacteria as cleansing agents is a useful (and often the only) approach to removing toxic chemicals from waters or soils contaminated by agricultural practices.

II. MICROBIAL DEGRADATION OF XENOBIOTIC MOLECULES

Microorganisms have amazing abilities when it comes to degradation of organic molecules. Virtually all naturally produced organic compounds, including natural products bearing unusual substituents such as halogen atoms, are degradable by some microbe or consortium of microbes. Some pure cultures of bacteria of the genus *Pseudomonas* can use more than 100 natural compounds as sole sources of carbon and energy. Fortunately, this catabolic versatility extends to many anthropogenic compounds as well. Microbes are not infallible, but they come close.

We will not attempt to review all of the known instances of microbial degradation of xenobiotic compounds. Such a compendium would be impressive and highly instructive, but would leave little opportunity to discuss the broader aspects of applying these catabolic abilities to pollution problems. Several recent reviews are available which tabulate specific man-made chemicals known to be degraded by microorganisms,[2,3] and the list grows longer each day.

For the purposes of the present discussions, we will summarize some of the most recent physiological and biochemical work being performed on the biodegradation of toxic, anthropogenic chemicals. We will discuss as much as possible chemicals of agricultural relevance that have a propensity to end up in aquatic systems and, finally, we will consider these chemicals in our succeeding discussion of pollution remediation technologies.

A. CHLORINATED PHENOLS

The chlorinated phenols comprise a large group of xenobiotic molecules used in agriculture for their pesticidal properties. Thousands of tons of chlorophenols are produced annually for use worldwide.[4,5] More than 50,000 tons of just one chlorophenol, pentachlorophenol (PCP), are produced yearly.[6] Chlorophenols are used as herbicides, fungicides, or general biocides, and since relatively few microbes can decompose them, chlorophenols released into the environment often persist long enough to become serious pollution problems

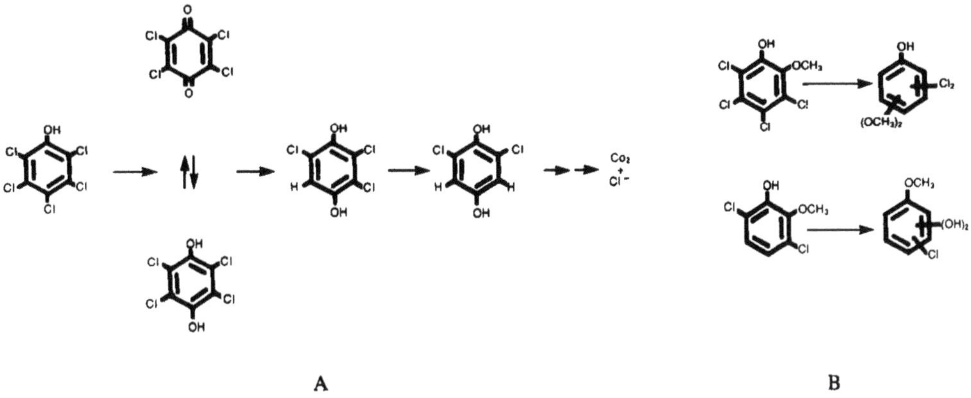

FIGURE 1. Bacterial catabolism of highly chlorinated phenols and guaiacols. (A) Pentachlorophenol;[14] (B) chloroguaiacols.[18]

affecting both soil and water. However, microbial physiologists have recently begun to understand the catabolic processes used by certain microbes to degrade chlorinated phenols, and biotechnologists have begun to apply those catabolic activities to remediation of chlorophenol pollution.[5]

PCP is degraded by a variety of microorganisms. A number of pure cultures of bacteria have been isolated that use PCP as a sole source of carbon and energy.[7-13] We, among others, recently have been examining the biochemistry of PCP degradation by some of these bacteria.[13-18] A nonclassical catabolic sequence involving hydrolytic and/or reductive dehalogenation of the PCP molecule seems to be involved, as shown in Figure 1A. These PCP degraders appear to funnel numerous polychlorinated phenols[5,17] and even polychloroguaiacols[18] (chlorohydroxymethoxybenzenes) into this novel pathway, as shown in Figure 1B. These will be very useful transformations for biotechnologists wishing to remove polychlorinated phenols or chloroguaiacols from contaminated natural environments (see Section III below).

Mono- or dichlorinated phenols appear to be degraded by bacteria via catabolic sequences that employ monooxygenases and dioxygenases.[19,20] These enzymes perform classic hydroxylation and ring-fission reactions long known to be important for microbial degradation of nonhalogenated aromatic molecules; however, the enzymes involved in catabolism of mono- and dichlorophenols often show specificities for substrates bearing halogens. Pathways known for the bacterial degradation of 4-chlorophenol and 2-chlorophenol are shown in Figure 2A and those for 2,4-dichlorophenol in Figure 2B.

It seems clear that each of the possible chlorophenols, from the three isomers of monochlorophenol through the numerous di-, tri-, and tetrachlorophenols and PCP, is usable by some bacterium as a sole source(s) of carbon and energy. As good biotechnologists, we should be able to apply these microbes to solving environmental problems relating to pollution by this class of toxic chemicals.

B. HALOALKANES

The most troubling class of xenobiotic molecules, from a standpoint of global scale of contamination, potential for adverse effects on human health, and difficulty in remediation once contamination has occurred, is that group that includes the many halogenated alkanes. These versatile compounds are produced in vast quantities worldwide. They are employed as industrial solvents, insecticides, preservatives, refrigerants, degreasing agents, and for a myriad of other uses throughout the industrialized world. Biotechnology holds great promise for solving pollution problems related to contamination of the environment by haloalkanes.

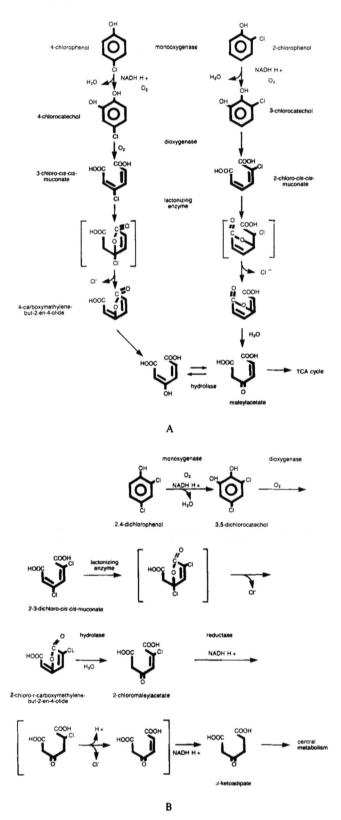

FIGURE 2.   Bacterial catabolism of monochloro- and dichlorophenols.
(A) 4-Chlorophenol;[19] (B) 2,4-dichlorophenol.[19,20]

Why? Because a surprising number of the most commonly encountered haloalkanes are biodegradable. We need only learn how to harness these biodegradative abilities of microbial communities for use by the developing bioremediation industry.

Some of the most recent literature reveals that microorganisms can degrade a wide spectrum of particularly troubling haloalkanes. For example, Jannsen et al.[21] isolated an *Acinetobacter* strain that was able to grow with and quantitatively release halide from 1,6-dichlorohexane, 1,9-dichlorononane, 1-chloropentane, 1-chlorobutane, 1-bromopentane, ethylbromide, and 1-iodopropane. The dehalogenating activity of the *Acinetobacter* appeared to be catalyzed by a halidohydrolase. Other recent reports confirm degradation, by pure microbial cultures, of dichloromethane (by a *Pseudomonas* and a *Hyphomicrobium*),[22-24] haloacetates (by a *Moraxella*),[25] D,L-2-halo acids (by a *Pseudomonas*),[26] trichloromethane (by a *Methylococcus*),[27] and trichloroethylene (by a bacterium identified tentatively as genus *Acinetobacter*).[28] In the latter instance, it appeared that the enzyme(s) responsible for degradation of trichloroethylene (completely to CO_2 and chloride) was actually an enzyme(s) induced for the degradation of aromatic molecules such as phenol, *o*-cresol, or *m*-cresol.[29] This is an example of what some investigators refer to as "co-metabolism" or "fortuitous metabolism". Whatever the phrase applied, it is a very useful attribute of which advantage can be taken to remove recalcitrant xenobiotic molecules from contaminated environments.

Degradation of haloalkanes also can be accomplished by mixtures of microorganisms, known as "consortia". For example, Pignatello[30] found that aerobic microbial consortia of two types of surface soils from an area receiving groundwater discharges contaminated with ethylene dibromide were able to mineralize ethylene dibromide to CO_2 and bromide. Vogel and McCarty[31] found that an anaerobic methanogenic microbial consortium was able to degrade tetrachloroethylene to CO_2, probably by way of trichloroethylene, dichloroethylene, and vinyl chloride. Complete mineralization of trichloroethylene was observed when non-sterile soils were exposed to natural gas.[32] Some of these activities of microbial consortia involve fortuitous metabolism of the chloroalkanes, but, nevertheless, such activities may be manipulated such that they may be employed for remediation of pollution.

C. NITROAROMATIC MOLECULES

Nitroaromatic molecules form a class of agricultural chemicals used primarily as herbicides (e.g., Dinoseb [2-*sec*-butyl-4,6-dinitrophenol], a contact herbicide) and insecticides (e.g., parathion). These types of compounds are widely used and very toxic. They are frequent soil and water contaminants in places such as pesticide formulation areas.

Spain et al.[33] isolated a *Moraxella* capable of growth on *p*-nitrophenol, which was degraded to CO_2, NO_2^-, and H_2O. This work confirmed an earlier investigation by Simpson and Evans[34] indicating the oxidative release of nitrite from *p*-nitrophenol by a soil bacterium. McCormick et al.[35] found that anaerobic bacteria such as *Veillonella alcalescens* could degrade nitroaromatic compounds by reductive mechanisms: R-NO_2 → R-NO → R-NHOH → R-NH_2. The aminoaromatic molecules formed are readily degraded by other soil microorganisms, with release of the nitrogen substituent as ammonia.[41] The enzymes that reduce aromatic nitro groups have been called "nitroreductases" and have been demonstrated in both aerobic and anaerobic microorganisms.[36,37]

Bacteria are also able to degrade nitrobenzoic acid,[38] as well as *o*- and *p*-nitrophenols.[39] As in the studies discussed above, the nitro group is released from the ring as nitrite. Zeyer et al.[40] characterized a nitrophenol oxygenase responsible for release of nitrite from *o*-nitrophenol and found that it was also active against 4-methyl-4-chloro- and 4-formyl-*o*-nitrophenols.

Experiments performed in natural environments confirm the biodegradability of nitroaromatic compounds. Hallas and Alexander[42] found that nitrobenzene, 3- and 4-nitrobenzoic acids, 3- and 4-nitrotoluenes, and 1,2- and 1,3-dinitrobenzenes disappeared from sewage

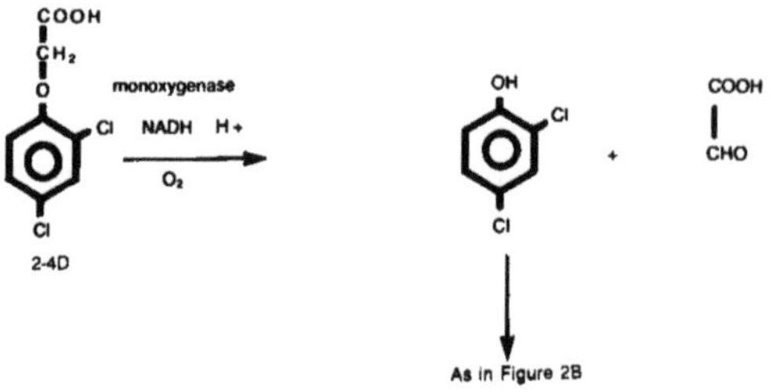

FIGURE 3.   Bacterial catabolism of 2,4-dichlorophenoxyacetic acid.

effluent in the presence or absence of oxygen. Aminoaromatic compounds appeared to be intermediates in the biodegradation process. Munnecke and Hsieh[43] cultivated in their laboratory a mixed culture of microorganisms that was able to grow on parathion. A pseudomonad was isolated from the mixture and was shown to grow on the parathion hydrolysis product *p*-nitrophenol. Likewise, others have confirmed the biodegradability of parathion in soils of various types.[44,45]

D. CHLORINATED BIPHENYLS

The polychlorinated biphenyls (PCBs, used in past years in vast quantities in high-temperature applications such as transformer oils) are among the most notorious of soil and water pollutants, yet even these are subject to considerable biodegradation. Monohalogenated biphenyls generally are readily mineralized by microbial communities[46] and pure microbial cultures.[47] We have isolated a *Pseudomonas* sp. that degrades both monochlorinated biphenyls and the monochlorobenzoic acids that in some instances are formed as chlorobiphenyl biotransformation products.[94] Interestingly, this pseudomonad does not grow well on unchlorinated biphenyl. Baxter et al.[48] found that *Nocardia* and *Pseudomonas* spp. could degrade the mono- through tetrachlorobiphenyls present in Aroclor® 1252. Mixed microbial cultures have been reported to degrade most of the mono-, di-, and trichlorobiphenyls and some tetrachlorobiphenyls in water saturated with Aroclor® 1242.[49] Bedard et al.[50] isolated a strain of *Alcaligenes eutrophus* from PCB-containing dredge spoils by enrichment on biphenyl. Their strain grew well on biphenyl and 2-chlorobiphenyl and under proper conditions rapidly degraded many tetra-, penta-, and several hexachlorobiphenyls. These authors concluded that a two-step process consisting of anaerobic dechlorination followed by oxidation by the *Alcaligenes* might result in degradation of all the congeners in the Aroclor® 1242 and possibly Aroclor® 1254.

E. CHLORINATED PHENOXY HERBICIDES

The chlorinated phenoxy herbicides are used to control broadleaf weeds in agricultural fields, forests, and lawns. They include compounds such as 2,4-dichlorophenoxyacetic acid (2,4-D), 2,4,5-trichlorophenoxyacetic acid (2,4,5-T), and analogs of these compounds. Under most environmental conditions these compounds are readily biodegradable, though the trichlorinated derivatives are more persistent than the dichloro derivatives. More problematic are the chlorinated polychloro-*p*-dibenzodioxins which have been associated as contaminants, particularly of the trichlorophenoxy herbicides.

The herbicide 2,4-D is readily degraded by numerous pure bacterial cultures.[51] The plasmid-encoded catabolic pathway thought to be employed by most bacteria that degrade 2,4-D is shown in Figure 3.[52] The compound 2,4,5-T is degradable by a *Pseudomonas cepacia* isolated and studied by Chakrabarty's group.[53] The pathway for this herbicide is still under study.

F. CHLOROBENZENES

Chlorobenzenes find uses as solvents, degreasers, and intermediates in the manufacture

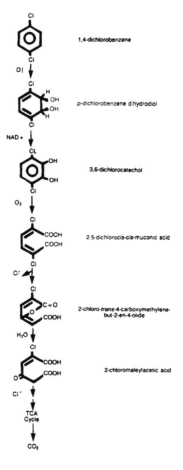

FIGURE 4. Bacterial catabolism of 1,4-dichlorobenzene.[58]

of dyes and pesticides. Evidence is conclusive that a variety of chlorobenzenes are degradable by microorganisms. Bouwer and McCarty observed mineralization of chlorobenzene, 1,4-dichlorobenzene, 1,2-dichlorobenzene, and 1,2,4-trichlorobenzene by a microbial consortium in the form of a microbial biofilm.[54] Other studies of microbial consortia confirm these abilities.[55,56] Pure cultures of *Pseudomonas*[57,58] and *Alcaligenes*[59] spp. have been shown to utilize dichlorobenzenes as sole sources of carbon and energy. Spain and Nishino[58] proposed a pathway for the degradation of 1,4-dichlorobenzene by their pseudomonad, as illustrated in Figure 4. The enzymes induced for degradation of the dichlorobenzene appeared to be specific for the chlorinated substrates, as enzymes of the *ortho* pathway induced by growth on benzene did not attack chlorobenzenes or chlorocatechols. This is an example of catabolic pathways specific to chlorinated substrates, as has been observed previously for other chloroaromatic molecules such as 5-chlorosalicylic acid.[60]

G. AROMATIC AND POLYNUCLEAR AROMATIC HYDROCARBONS (PAHs)

Benzene and related compounds form a group of stable compounds containing only carbon and hydrogen in the form of a resonance-stabilized aromatic nucleus or nuclei. These compounds are found in nature as components of fossil fuels, and they are formed during pyrolysis of organic materials. A number of these compounds are highly useful solvents (e.g., benzene, toluene, and the xylenes) and/or precursors for the synthesis of industrial chemicals and pesticides. Thus, they may be considered important agriculturally oriented chemicals. Structures of some representative aromatic and polynuclear aromatic hydrocar-

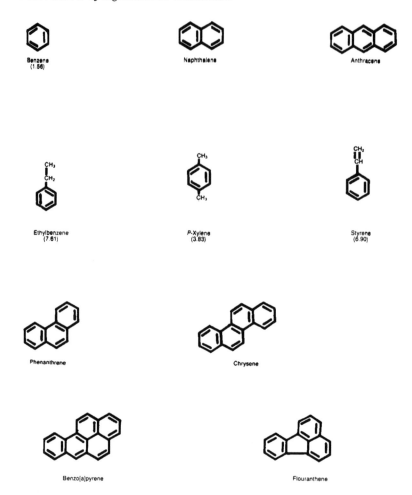

FIGURE 5.   Some representative aromatic and polynuclear aromatic hydrocarbons (1980 annual production in billions of pounds shown in parentheses).[61]

bons are shown in Figure 5. Many of these compounds or their oxidation products are carcinogenic.[61]

Only a small fraction of all known aromatic hydrocarbons have been examined for their biodegradability; however, it appears that as a class these molecules are degradable by microorganisms. At least the following representative molecules are known to be degraded or transformed by bacteria: benzene, toluene, ethylbenzene, *p*-, *m*-, and *o*-xylene, biphenyl, naphthalene, fluorene, phenanthrene, anthracene, benzo[*a*]anthracene, benzo[*a*]pyrene, and the dibenzanthracenes.[61] Aerobic biodegradation of aromatic hydrocarbons by bacteria usually involves introduction of oxygen into the substrate. This usually is initiated by an oxygenase that attacks the aromatic nucleus to form a dihydrodiol, with subsequent dehydrogenation of the dihydrodiol to yield a catechol. The catechol then is dearomatized by introduction of dioxygen into the ring, either between the hydroxyls or adjacent to them (Figure 6). If the hydrocarbon bears side chains, as with xylenes, then biodegradation may proceed either by initial attack on the ring as with benzene (Figure 7A) or by oxidation of the side chain to form an aromatic acid which is degraded further by other standard pathways (Figure 7B). Although there are many more steps in the bacterial degradation of polynuclear aromatic hydrocarbons than for single-ring molecules, the principles of oxidative degradation are similar, as is seen in the biodegradation of compounds such as phenanthrene (Figure 8).

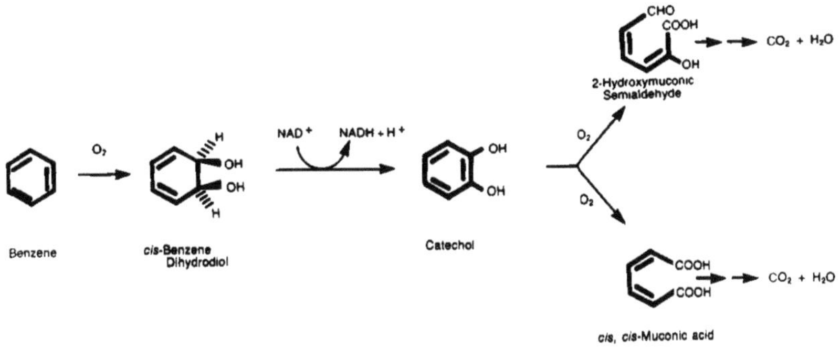

81

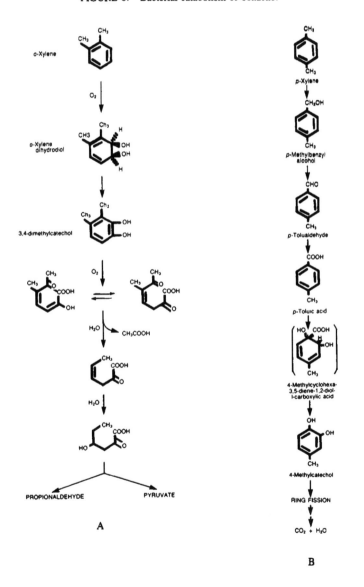

FIGURE 6.   Bacterial catabolism of benzene.[61]

FIGURE 7.   Bacterial catabolism of xylenes.[61] (A) Dihydroxylation route; (B) side-chain oxidation route.

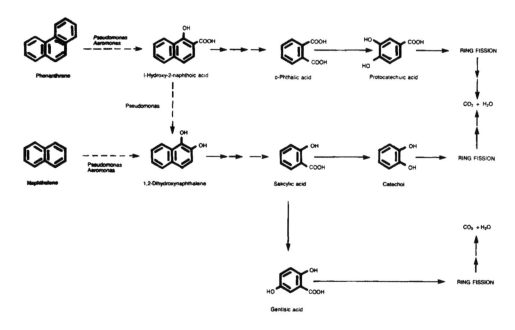

FIGURE 8.   Bacterial catabolism of phenanthrene.[61]

H. MISCELLANEOUS AGRICULTURAL CHEMICALS

There are a large number of pesticides used in agriculture that have complex structures that place them in special molecular subgroups (e.g., the triazines, organophosphorous compounds, carbamates, anilines, pyridines, etc.). Rather than discuss each of these groups in depth, we will give examples of bacterial degradation and/or transformation of some representative molecules. The point to be made is that even many of these complex, highly toxic agricultural chemicals are subject to extensive or total degradation by bacteria.

It has been shown that compounds such as s-ethyl-*N,N*-dipropylthiocarbamate will support the growth of bacteria. For example, an *Arthrobacter* was isolated using this compound as a sole carbon/energy source.[62] The degradative pathway appeared to be encoded on a 50.5-MDa plasmid.

Various chloroaniline pesticides are degraded by bacteria. A strain of *Pseudomonas putida* was isolated on 3,4-dichloroaniline and shown to degrade this compound via dichlorocatechol.[63] Another pseudomonad was shown to use 4-chloroaniline as both a carbon and a nitrogen source.[64]

Diuron (3,[3,4-dichlorophenyl]1,1-dimethylurea) has been shown to be degradable by an aerobic microbial consortium,[65] xanthone (a multiring compound having structural similarities to the highly toxic dibenzodioxins and dibenzofurans) by an *Arthrobacter*[66] (Figure 9), atrazine (2-chloro-4-ethylamino-6-[isopropylamino]-s-triazine) by a pure culture of *Pseudomonas*,[67-69] metamitron (3-methyl-4-amino-6-1,2,4-5[4H]-one) by an *Arthrobacter*,[70] and glyphosate (the active ingredient in the herbicide Roundup®) by a variety of microbes.[71-73]

This type of listing could go on and on; however, the point we stress has been made clear: bacteria are excellent degraders of agricultural chemicals. We need to learn how to harness these biodegradative abilities for the cleanup of polluted soils and waters.

III. USE OF BACTERIA TO CLEANSE POLLUTED SOILS AND WATERS

The use of bacteria to detoxify soils and waters contaminated by exotic agricultural and industrial chemicals is a young, developing technology. However, there already are a number

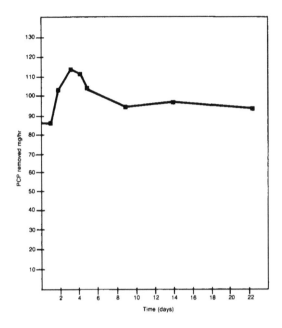

FIGURE 9. Degradation of pentachlorophenol by *Flavobacterium* cells immobilized in alginate beads. Cells (7 g wet weight) were immobilized in 200 ml of 2% alginate in the form of ~2 mm beads, fluidized with air in a 60 cm × 1 in. column topped by a 1-l reservoir; influent (PCP) = 670 ppm, effluent (PCP) not detectable by measuring absorbance at 320 nm (PCP, λ_{max}); total volume of fluidized bed = 400 ml. (From Steiert, J. G., Ph.D. thesis, University of Minnesota, Minneapolis, 1987. With permission.)

of new, biotechnology-oriented companies which are making this approach to the remediation of pollution their major emphasis. Though much of the technology being employed is of a proprietary nature, we can learn a fair amount from the published literature concerning the considerable promise of this biotechnological approach.

A. DETOXIFICATION OF SOIL BY DIRECT INOCULATION

There have been a number of successful attempts to remove toxic chemicals from soil by direct inoculation of contaminated soils with pollutant-degrading bacteria. Kilbane et al.[74] and Chatterjee et al.[75] removed the herbicide 2,4,5-T from soil by inoculation of contaminated soil with cells of a 2,4,5-T-degrading *Pseudomonas cepacia* strain. Edgehill and Finn[76] and Crawford and Mohn[77] demonstrated the removal of the general biocide pentachlorophenol from soils by inoculation with an *Arthrobacter* and a *Flavobacterium*, respectively. Biodegradation of parathion in contaminated soils was greatly enhanced by inoculation of the soils with parathion-degrading bacteria.[78] In all of these instances, microbial cells were grown in culture and added to the soils at high density (e.g., 10^6 to 10^8 cells per gram), an approach amenable to industrial operations where cells could be cultivated in great quantity using large fermentors, preserved (e.g., by freezing), and shipped to their site of use.

B. DETOXIFICATION OF SOIL BY MICROFLORA STIMULATION

It is possible to stimulate microorganisms already present in the soil to degrade a contaminating xenobiotic chemical. Initially, the microbes having the desired degradative abilities may be present in soil in very low numbers. What is needed is to stimulate that subpopulation to increase in numbers and to induce the proper suite(s) of enzymes in those new cells. An approach that often accomplishes these goals is termed "analog enrichment".

The concept of analog enrichment is logical and straightforward in its application. In the original work in this area by Focht and Alexander[79] and Horvath,[80] the concept was developed around the assumption that it should be possible to stimulate growth and enzyme induction in pure bacterial strains or bacteria in soil and water that are able to degrade a compound structurally similar to a particular xenobiotic molecule. This is accomplished by the addition of that structural "analog" to culture media (pure cultures) or the environment (soil or water). Then, hopefully, the stimulated microorganisms might degrade the target xenobiotic molecule fortuitously by the activities of enzymes lacking strict substrate specificities. This phenomenon has been called "co-metabolism", though it does not always fit the strict definition of co-metabolism as coined by Horvath,[80] where the analog does not yield sufficient energy for growth of the analog-degrading microbe.

Analog enrichment does appear to be an effective soil remediation technique. When aniline was added to soils contaminated by 3,4-dichloroaniline, stimulation of aniline degraders resulted in removal of the dichloroaniline.[81] Brunner et al.[82] significantly stimulated the removal of Aroclor® 1242 (a PCB containing 42% chlorine) from soils by adding unchlorinated biphenyl as an analog. By adding both biphenyl and a biphenyl-degrading *Acinetobacter* to PCB-contaminated soils, these investigators stimulated Aroclor® removal even more (20 to 27% mineralization to CO_2 after 210 d, as opposed to less than 3% mineralization in nonstimulated controls). In later work, Focht's group increased Aroclor® 1242 removal by activated sludges to essentially 100% by analog enrichment with certain mono- and dichlorobenzoic acids.[95]

C. DETOXIFICATION OF WATER

Direct inoculation of natural waters with pollutant-degrading bacteria is an even more effective decontamination technique than direct inoculation of soils. For example, Martinson et al.[83] added cells of a pentachlorophenol-degrading *Flavobacterium* to a variety of penta-contaminated natural waters, including waters from streams and lakes and groundwaters. Removal of the penta (100 ppm) was fast and complete, and addition of the bacteria was sufficient to protect fathead minnows from being killed by this highly lethal level of toxicant.

A highly effective way to decontaminate water containing xenobiotic molecules is to pass the contaminated water through a reactor packed with immobilized, pollutant-degrading bacterial cells. This approach is especially convenient since bacteria can be grown in large quantity by use of fermentors, concentrated by filtration or centrifugation, and entrapped in beads (or other packing configurations) where they often are long-lived and express excellent xenobiotic compound-degrading activities.

Rehm's group has examined the degradation of phenol[85] and 4-chlorophenol[86] by immobilized cells. These investigators used strains of *Alicaligenes*, *Candida* (not a bacterium, but handled in much the same way one would handle bacteria), and *Pseudomonas*, immobilizing them in matrices including alginate, polyacrylamide, and activated carbon. Immobilized cells degraded the phenolic compounds at higher concentrations and at faster rates than did free cells. Although data were not presented to support their hypothesis, the authors suggested that high cell densities in the immobilization matrices might have served to protect the cells by lowering the local concentration of the toxic substances.

Bettmann and Rehm[85] described a continuous-flow column bioreactor containing cells of *Pseudomonas putida* immobilized within polyacrylamide. The column had a working volume of 900 ml, contained 100 g of beads, and had a flow rate of 1 ml/min (dilution rate = 0.67 h^{-1}). Under sterile conditions with phenol as the only carbon source, a degradation rate of 7.2 g phenol per liter per day was observed over a 400-h period. This represents more than 2.5 g of phenol degraded by this small bioreactor in slightly over 2 weeks. Such degradation rates are impressive, as is the stability of such a system. When nonsterile wastewater was used, the degradation rate fell to 3.12 g/l/d, still a significant rate.

Klein et al.[88] investigated the kinetics of phenol degradation by free and immobilized cells of *Candida tropicalis,* finding zero-order kinetics with both systems when the phenol concentration was <1 g/l. The availability of oxygen to the cells was a major factor controlling degradation rates.

The kinetics of phenol degradation have been examined using an immobilized methanogenic microbial consortium.[89] The immobilized cells tolerated higher concentrations of phenol than did the free cells.

A benzene-degrading *Pseudomonas putida* strain was immobilized in polyacrylamide.[90] Although the cells lost up to 70% of their activity upon immobilization, the activity was restored by incubating the cells in a medium containing benzene and succinate. Once reactivated, the cells retained their full activity, within a column, for 73 d.

In our laboratory we have immobilized several pollutant-degrading bacteria in supports such as alginate. For example, we immobilized a PCP-degrading *Flavobacterium* and studied its degradation in small-batch, fluidized-bed, and packed-column reactors. One of our fluidized-bed reactors was able to reduce influent penta concentrations as high as 670 ppm to undetectable levels, and it maintained this level of activity for at least 22 d (Figure 9).

We have also examined the kinetics of degradation of cresols (methylphenols) by immobilized bacterial cells. A *p*-cresol-degrading bacterium was immobilized in alginate beads and the beads used in small-batch reactors with concentrations as high as 1000 ppm. The rate of degradation was dependent on the method of aeration, with the greatest rate detected in oxygen-gas, air-lift reactors.

Thus, immobilization of xenobiotic compound-degrading bacteria shows great promise as a water pollution remediation technology. Immobilized cells often are able to tolerate concentrations of toxic chemicals that will kill free cells. Usually there is long-term stability of degradative activity, and inactive matrices often can be reactivated by incubation under appropriate conditions. The ability to maintain a very high biomass density in reactors leads to reactors having very high total activities (g pollutant removed $\times$ reactor volume^{-1} $\times$ unit time^{-1}). A variety of immobilization supports can be employed, including polysaccharides (alginate, carrageenan, etc.)[91] and polyurethane foams.[92,93] Numerous reactor configurations are possible, including batch reactors, fixed-bed (column type) reactors, and fluidized-bed reactors. Finally, immobilized cell bioreactors may be customized to treat particular types of contaminated waters by mixing supports containing different pollutant-degrading bacteria immobilized separately or together.

IV. SUMMARY

Bacteria are ideal waste removers. Bacteria are not infallible, but they come close. They are able to degrade (usually to completely mineralized products) virtually all of the types of xenobiotic molecules employed in modern agriculture. As microbiologists and microbial ecologists, we can learn to harness these catabolic abilities and use them to clean our environment of toxic chemicals. Cooperation between microbiologists and engineers is leading to rapid advances in techniques to apply bacterial processes to the remediation of soil and water pollution. Bacteria will play a major role in future efforts to rid our environment of hazardous chemicals.

ACKNOWLEDGMENTS

Our work discussed here was supported in part by the Idaho Agricultural Experiment Station, University of Idaho, and the U.S. Public Health Service under grant ESO3270-02 from the National Institute of Environmental Health Sciences. We thank Jack Steiert for allowing use of data from his Ph.D. thesis.

REFERENCES

1. **McCormick, D.**, One bug's meat . . . , *Biotechnology*, 3, 429, 1985.
2. **Kobayashi, H. and Rittmann, B. E.**, Microbial removal of hazardous organic compounds, *Environ. Sci. Technol.*, 16, 170A, 1982.
3. **Peyton, T. O.**, Biological disposal of hazardous waste, *Enzyme Microb. Technol.*, 6, 146, 1984.
4. **Reineke, W.**, Microbial degradation of halogenated aromatic compounds, in *Microbial Degradation of Organic Compounds*, Gibson, D. T., Ed., Marcel Dekker, New York, 1984, 319.
5. **Steiert, J. G. and Crawford, R. L.**, Microbial degradation of chlorinated phenols, *Trends Biotechnol.*, 3, 300, 1985.
6. **Crosby, D. G.**, Environmental chemistry of pentachlorophenol, *Pure Appl. Chem.*, 53, 1051, 1981.
7. **Chu, J. P. and Kirsch, E. J.**, Metabolism of pentachlorophenol by an axenic bacterial culture, *Appl. Microbiol.*, 23, 1033, 1972.
8. **Watanabe, I.**, Isolation of pentachlorophenol decomposing bacteria from soil, *Soil Sci. Plant Nutr.*, 19, 109, 1973.
9. **Edgehill, R. U. and Finn, R. K.**, Isolation, characterization and growth kinetics of bacteria metabolizing pentachlorophenol, *Eur. J. Appl. Microbiol. Biotechnol.*, 16, 179, 1982.
10. **Stanlake, G. J. and Finn, R. K.**, Isolation and characterization of a pentachlorophenol-degrading bacterium, *Appl. Environ. Microbiol.*, 44, 1421, 1982.
11. **Suzuki, T.**, Metabolism of pentachlorophenol (PCP) by a soil microorganism, *Nogyo Gijutsu Kenkyusho Hokoku C*, 38, 69, 1983.
12. **Saber, D. L. and Crawford, R. L.**, Isolation and characterization of *Flavobacterium* strains that degrade pentachlorophenol, *Appl. Environ. Microbiol.*, 50, 1512, 1985.
13. **Apajalahti, J. H. A. and Salkinoja-Salonen, M. S.**, Degradation of polychlorinated phenols by *Rhodococcus chlorophenolicus*, *Appl. Microbiol. Biotechnol.*, 25, 62, 1986.
14. **Steiert, J. G. and Crawford, R. L.**, Catabolism of pentachlorophenol by a *Flavobacterium* sp., *Biochem. Biophys. Res. Commun.*, 141, 825, 1986.
15. **Reiner, E. A., Chu, J., and Kirsch, E. J.**, Microbial metabolism of pentachlorophenol, in *Pentachlorophenol*, Rao, K. R., Ed., Plenum Press, New York, 1978, 67.
16. **Suzuki, T.**, Metabolism of pentachlorophenol by a soil microbe, *J. Environ. Sci. Health*, B12, 113, 1977.
17. **Apajalahti, J. H. A. and Salkinoja-Salonen, M. S.**, Dechlorination and *para*-hydroxylation of polychlorinated phenols by *Rhodococcus chlorophenolicus*, *J. Bacteriol.*, 169, 675, 1987.
18. **Haggblom, M., Apajalahti, J., and Salkinoja-Salonen, M.**, Metabolism of chloroguaiacols by *Rhodococcus chlorophenolicus*, *Appl. Microbiol. Biotechnol.*, 24, 397, 1986.
19. **Knackmuss, H.-J.**, Degradation of halogenated and sulfonated hydrocarbons, in *Microbial Degradation of Xenobiotics and Recalcitrant Compounds*, Leisinger, T., Hutter, R., Cook, A. M., and Nuesch, J., Eds., Academic Press, New York, 1981, 189.
20. **Chapman, P. J.**, Degradation mechanisms, in Proc. Workshop Microbial Degradation of Pollutants in Marine Environments, Bourquin, A. W. and Pritchard, P. H., Eds., U.S. Environmental Protection Agency, Washington, D.C., 1978, 28.
21. **Janssen, D. B., Jager, D., and Witholt, B.**, Degradation of n-haloalkanes and α-ω-dihaloalkanes by wild-type and mutants of *Acinetobacter* sp. strain GJ70, *Appl. Environ. Microbiol.*, 53, 561, 1987.
22. **Kohler-Staub, D. and Leisinger, T.**, Dichloromethane dehalogenase of *Hyphomicrobium* sp. strain DM2, *J. Bacteriol.*, 162, 676, 1985.
23. **Brunner, W., Staub, D., and Leisinger, T.**, Bacterial degradation of dichloromethane, *Appl. Environ. Microbiol.*, 40, 950, 1980.
24. **Stucki, G., Gälli, R., Ebersold, H.-R., and Leisinger, T.**, Dehalogenation of dichloromethane by cell extracts of *Hyphomicrobium* DM2, *Arch. Microbiol.*, 130, 366, 1981.
25. **Kawasaki, H., Tone, N., and Tonomura, K.**, Plasmid-determined dehalogenation of haloacetate in *Moraxella* species, *Agric. Biol. Chem.*, 45, 29, 1981.
26. **Motosugi, K., Esaki, N., and Soda, K.**, Purification and properties of a new enzyme, D,L-2-haloacid dehalogenase, from *Pseudomonas* sp., *J. Bacteriol.*, 150, 522, 1982.
27. **Dalton, H. and Stirling, D. I.**, Co-metabolism, *Philos. Trans. R. Soc. London Ser. B*, 297, 481, 1982.
28. **Nelson, M. J. K., Montgomery, S. O., O'Neill, E. J., and Pritchard, P. H.**, Aerobic metabolism of trichloroethylene by a bacterial isolate, *Appl. Environ. Microbiol.*, 52, 383, 1986.
29. **Nelson, M. J. K., Montgomery, S. O., Mahaffey, W. R., and Pritchard, P. H.**, Biodegradation of trichloroethylene and involvement of an aromatic biodegradative pathway, *Appl. Environ. Microbiol.*, 53, 949, 1987.
30. **Pignatello, J. J.**, Ethylene dibromide mineralization in soils under aerobic conditions, *Appl. Environ. Microbiol.*, 51, 588, 1986.

31. **Vogel, T. M. and McCarty, P. L.**, Biotransformation of tetrachloroethylene to trichloroethylene, dichloroethylene, vinyl chloride, and carbon dioxide under methanogenic conditions, *Appl. Environ. Microbiol.*, 49, 1080, 1985.

32. **Wilson, J. T. and Wilson, B. H.**, Biotransformation of trichloroethylene in soil, *Appl. Environ. Microbiol.*, 49, 242, 1985.

33. **Spain, J. C., Wyss, O., and Gibson, D. T.**, Enzymatic oxidation of *p*-nitrophenol, *Biochem. Biophys. Res. Commun.*, 88, 634, 1979.

34. **Simpson, J. R. and Evans, W. C.**, The metabolism of nitrophenols by certain bacteria, *Biochem. J.*, 55, 24, 1953.

35. **McCormick, N. G., Feeherry, F. E., and Levinson, H. S.**, Microbial transformation of 2,4,6-trinitrotoluene and other nitroaromatic compounds, *Appl. Environ. Microbiol.*, 31, 949, 1976.

36. **Kinouchi, T. and Ohnishi, Y.**, Purification and characterization of 1-nitropyrene nitroreductases from *Bacteroides fragilis*, *Appl. Environ. Microbiol.*, 46, 596, 1983.

37. **Villanueva, J. R.**, The purification of a nitro-reductase from *Nocardia* V, *J. Biol. Chem.*, 239, 773, 1965.

38. **Cartwright, N. J. and Cain, R. B.**, Bacterial degradation of nitrobenzoic acids, *Biochem. J.*, 71, 248, 1959.

39. **Zeyer, J. and Kearney, P. C.**, Degradation of *o*-nitrophenol and *m*-nitrophenol by a *Pseudomonas putida*, *J. Agric. Food Chem.*, 32, 238, 1984.

40. **Zeyer, J., Kocher, H. P., and Timmis, K. N.**, Influence of *para*-substituents on the oxidative metabolism of *o*-nitrophenol by *Pseudomonas putida* B2, *Appl. Environ. Microbiol.*, 52, 33, 1986.

41. **Parris, G. E.**, Environmental and metabolic transformations of primary aromatic amines and related compounds, *Residue Rev.*, 76, 1, 1980.

42. **Hallas, L. and Alexander, M.**, Microbial transformation of nitroaromatic compounds in sewage effluent, *Appl. Environ. Microbiol.*, 45,1234,1983

43. **Munnecke, D. M. and Hsieh, D. P. H.**, Microbial decontamination of parathion and *p*-nitrophenol in aqueous media, *Appl. Microbiol.*, 28, 212, 1974.

44. **Lichtenstein, E. P. and Schultz, K. R.**, The effects of moisture and microorganisms on the persistence and metabolism of some organophosphorus insecticides in soils, with special emphasis on parathion, *J. Econ. Entomol.*, 57, 618, 1964.

45. **Sethunathan, N.**, Degradation of parathion in flooded acid soils, *J. Agric. Food Chem.*, 21, 602, 1973.

46. **Kong, H. L. and Sayler, G. S.**, Degradation and total mineralization of monohalogenated biphenyls in natural sediment and mixed bacterial culture, *Appl. Environ. Microbiol.*, 46, 666, 1983.

47. **Furukawa, K., Tomizuka, N., and Kamabayashi, A.**, Effect of chlorine substitution on the bacterial metabolism of various polychlorinated biphenyls, *Appl. Environ. Microbiol.*, 38, 301, 1979.

48. **Baxter, R. A., Gilbert, R. E., Lidgett, R. A., Mainprize, J. H., and Voden, H. A.**, The degradation of polychlorinated biphenyls by microorganisms, *Sci. Total Environ.*, 4, 53, 1975.

49. **Clark, R. R., Chian, E. S. K., and Griffin, R. A.**, Degradation of polychlorinated biphenyls by mixed microbial cultures, *Appl. Environ. Microbiol.*, 37, 680, 1979.

50. **Bedard, D. L., Wagner, R. E., Brennan, M. J., Haberl, M. L., and Brown, J. F., Jr.**, Extensive degradation of Aroclors and environmentally transformed polychlorinated biphenyls by *Alcaligenes eutrophus* H850, *Appl. Environ. Microbiol.*, 53, 1094, 1987.

51. **Evans, W. C., Smith, B. S. W., Fernley, H. N., and Davies, J. I.**, Bacterial metabolism of 2,4-dichlorophenoxyacetate, *Biochem. J.*, 122, 543, 1971.

52. **Chapman, P. J.**, Degradation mechanisms, in Proc. Workshop Microbial Degradation of Pollutants in Marine Environments, Bourquin, A. W. and Pritchard, P. H., Eds., U.S. Environmental Protection Agency, Washington, D.C., 1978, 28.

53. **Kilbane, J. J., Chatterjee, D. K., Karns, J. S., Kellogg, S. T., and Chakrabarty, A. M.**, Biodegradation of 2,4,5-trichlorophenoxyacetic acid by a pure culture of *Pseudomonas cepacia*, *Appl. Environ. Microbiol.*, 44, 72, 1982.

54. **Bouwer, E. J. and McCarty, P. L.**, Removal of trace chlorinated organic compounds by activated carbon and fixed film bacteria, *Environ. Sci. Technol.*, 16, 836, 1982.

55. **Schwarzenbach, R. P., Ginger, W., Hoehn, E., and Schneider, J. K.**, Behavior of organic compounds during infiltration of river water to groundwater: field studies, *Environ. Sci. Technol.*, 17, 472, 1983.

56. **Kuhn, E. P., Colberg, P., Schnoor, L., Wanner, O., Zehnder, A. J. B., and Schwarzenbach, R. P.**, Microbial transformation of substituted benzenes during infiltration of river water to groundwater: laboratory column studies, *Environ. Sci. Technol.*, 19, 961, 1985.

57. **Reineke, W. and Knackmuss, H.-J.**, Hybrid pathway for chlorobenzene metabolism in *Pseudomonas* sp. B13 derivatives, *J. Bacteriol.*, 142, 467, 1980.

58. **Spain, J. C. and Nishino, S. F.**, Degradation of 1,4-dichlorobenzene by a *Pseudomonas* sp., *Appl. Environ. Microbiol.*, 53, 1010, 1987.

59. **deBont, J. A. M., Vorage, M. J. A., Hartmans, S., and van den Tweel, W. J. J.**, Microbial degradation of 1,3-dichlorobenzene, *Appl. Environ. Microbiol.*, 52, 677, 1986.

60. **Crawford, R. L., Olsen, P. E., and Frick, T. D.**, Catabolism of 5-chlorosalicylate by a *Bacillus* isolated from the Mississippi River, *Appl. Environ. Microbiol.*, 38, 379, 1979.
61. **Gibson, D. T. and Subramanian, V.**, Microbial degradation of aromatic hydrocarbons, in *Microbial Degradation of Organic Compounds*, Gibson, D. T., Ed., Marcel Dekker, New York, 1984, 181.
62. **Tam, A. C., Behki, R. M., and Shahamat, U. K.**, Isolation and characterization of an s-ethyl-N,N-dipropylthiocarbamate-degrading *Arthrobacter* strain and evidence for plasmid-associated s-ethyl-N,N-dipropylthiocarbamate degradation, *Appl. Environ. Microbiol.*, 53, 1088, 1987.
63. **You, I. S. and Bartha, R.**, Metabolism of 3,4-dichloroaniline by *Pseudomonas putida*, *J. Agric. Food Chem.*, 30, 274, 1982.
64. **Zeyer, J. and Kearney, P. C.**, Microbial degradation of *para*-chloroaniline as sole carbon and nitrogen source, *Pestic. Biochem. Physiol.*, 17, 215, 1982.
65. **Ellis, P. A. and Camper, N. D.**, Aerobic degradation of diuron by aquatic microorganisms, *J. Environ. Health*, B17, 277, 1982.
66. **Tomasek, P. H. and Crawford, R. L.**, Initial reactions of xanthone biodegradation by an *Arthrobacter* sp., *J. Bacteriol.*, 167, 818, 1986.
67. **Cook, A. and Hutter, R.**, s-Triazines as nitrogen sources for bacteria, *J. Agric. Food Chem.*, 29, 1135, 1981.
68. **Behki, R. and Kahn, S.**, Degradation of atrazine by *Pseudomonas:* N-dealkylation and dehalogenation of atrazine and its metabolites, *J. Agric. Food Chem.*, 43, 746, 1986.
69. **Cook, A. and Hutter, R.**, Diethylsimazine: bacterial dechlorination, deamination and complete degradation, *J. Agric. Food Chem.*, 32, 581, 1984.
70. **Englehardt, G., Ziegler, W., Wallnofer, P., Jaroczyk, H., and Oehlmann, L.**, Degradation of the triazinone herbicide metamitron by *Arthrobacter* sp. DSM20389, *J. Agric. Food Chem.*, 30, 278, 1982.
71. **Shinabarger, D. L. and Braymer, H. D.**, Glyphosate catabolism by *Pseudomonas* sp. strain PG2982, *J. Bacteriol.*, 168, 702, 1986.
72. **Jacobs, G., Schaefer, J., Stejskall, E., and McKay, R.**, Solid-state NMR determination of glyphosate metabolism in a *Pseudomonas* sp., *J. Biol. Chem.*, 260, 5899, 1985.
73. **La Nauze, J. M., Coggins, J. R., and Dixon, H. B. F.**, Aldolase-like imine formation in the mechanism of action of phosphoacetaldehyde hydrolase, *Biochem. J.*, 165, 409, 1977.
74. **Kilbane, J. J., Chatterjee, D. K., and Chakrabarty, A. M.**, Detoxification of 2,4,5-trichlorophenoxyacetic acid from contaminated soil by *Pseudomonas cepacia*, *Appl. Environ. Microbiol.*, 45, 1697, 1983.
75. **Chatterjee, D. K., Kilbane, J. J., and Chakrabarty, A. M.**, Biodegradation of 2,4,5-trichlorophenoxyacetic acid in soil by a pure culture of *Pseudomonas cepacia*, *Appl. Environ. Microbiol.*, 44, 514, 1982.
76. **Edgehill, R. U. and Finn, R. K.**, Microbial treatment of soil to remove pentachlorophenol, *Appl. Environ. Microbiol.*, 45, 1122, 1983.
77. **Crawford, R. L. and Mohn, W. W.**, Microbiological removal of pentachlorophenol from soil using a *Flavobacterium*, *Enzyme Microb. Technol.*, 7, 617, 1985.
78. **Barles, R. W., Daughton, C. G., and Hsieh, D. P. H.**, Accelerated parathion degradation in soil inoculated with acclimated bacteria under field conditions, *Arch. Environ. Contam. Toxicol.*, 8, 647, 1979.
79. **Focht, D. D. and Alexander, M.**, DDT metabolites and analogs: ring fission by *Hydrogenomonas*, *Science*, 170, 91, 1970.
80. **Horvath, R. S.**, Microbial co-metabolism and the degradation of organic compounds in nature, *Bacteriol. Rev.*, 36, 146, 1972.
81. **You, I. S. and Bartha, R.**, Stimulation of 3,4-dichloroaniline mineralization by aniline, *Appl. Environ. Microbiol.*, 44, 678, 1982.
82. **Brunner, W., Sutherland, F. H., and Focht, D. D.**, Enhanced biodegradation of polychlorinated biphenyls in soil by analog enrichment and bacterial inoculation, *J. Environ. Qual.*, 14, 324, 1985.
83. **Martinson, M. M., Steiert, J. G., Saber, D. L., and Crawford, R. L.**, Microbiological decontamination of pentachlorophenol in natural waters, in *Proc. 6th Int. Biodeterioration Symp.*, Commonwealth Agricultural Bureaux, London, 1985, 529.
84. **Bettmann, H. and Rehm, H.**, Degradation of phenol by polymer entrapped microorganisms, *Appl. Microbiol. Biotechnol.*, 20, 285, 1984.
85. **Bettmann, H. and Rehm, H.**, Continuous degradation of phenol(s) by *Pseudomonas putida* P8 entrapped in polyacrylamide-hydrazine, *Appl. Microbiol. Biotechnol.*, 22, 389, 1985.
86. **Westmeier, F. and Rehm, H.**, Biodegradation of 4-chlorophenol by entrapped *Alcaligenes* sp. A7-2, *Appl. Microbiol. Biotechnol.*, 22, 301, 1985.
87. **Westmeier, F. and Rehm, H.**, Degradation of 4-chlorophenol in municipal waste water by adsorptive immobilized *Alcaligenes* sp. A7-2, *Appl. Microbiol. Biotechnol.*, 26, 70, 1987.
88. **Klein, J., Hackel, U., and Wagner, F.**, Phenol degradation by *Candida tropicalis* whole cells entrapped in polymeric ionic network, in *Immobilized Microbial Cells*, Venkapsubramanin, K., Ed., American Chemical Society, Washington, D.C., 1979, 101.

89. **Dwyer, D., Krumme, M., Boyd, S., and Tiedje, J.,** Kinetics of phenol biodegradation by an immobilized methanogenic consortium, *Appl. Environ. Microbiol.,* 52, 345, 1986.

90. **Summerville, H., Mason, J., and Ruffell, R.,** Benzene degradation by bacterial cells immobilized in polyacrylamide gel, *Eur. J. Appl. Microbiol.,* 41, 75, 1977.

91. **Tosa, T.,** Use of immobilized cells, *Annu. Rev. Biophys. Bioeng.,* 10, 197, 1981.

92. **Fukui, S., Sonomoto, K., Itoh, N., and Tanaka, A.,** Several novel methods for immobilization of enzymes, microbial cells and organelles, *Biochemie,* 62, 381, 1980.

93. **Klein, J. and Kluge, M.,** Immobilization of microbial cells in polyurethane matrices, *Biotechnol. Lett.,* 3, 65, 1981.

94. **Barton, M. and Crawford, R.,** Novel biotransformations of 4-chlorobiphenyl by a *Pseudomonas* sp., *Appl. Environ. Microbiol.,* 54, 594, 1988.

95. **Focht, D. D. et al.,** personal communication.

Chapter 7

ALGAE AS IDEAL WASTE REMOVERS: BIOCHEMICAL PATHWAYS

Donald G. Redalje, Eirik O. Duerr, Joël de la Noüe, Patrick Mayzaud,
Arthur M. Nonomura, and Richard Cassin

TABLE OF CONTENTS

I. INTRODUCTION

Microalgae have often been considered ideal waste removers for sewage effluents because of their requirements for dissolved forms of both nitrogen and phosphorus, which are major components of wastewater. Another useful characteristic of microalgae is that they produce extracellular organic material which can then bind with dissolved metals, thereby reducing or eliminating metal toxicity. As with all particles in a fluid medium, the cell surfaces of algae provide sites for the adsorption of dissolved organic materials and metals. Each of the above properties of microalgae makes them potentially useful for the removal of a wide variety of waste products found in agricultural wastewaters.

In this chapter we will examine the important biochemical pathways which make algae useful in the reclamation of agricultural wastewaters, with particular reference to dissolved inorganic and organic nitrogen and phosphorus and dissolved organic materials, including a wide variety of xenobiotic compounds and trace metals. We will also examine some of the recent advances in biotechnology which may result in the enhancement of the capacity of microalgae to remove certain components of agricultural wastewater.

Microalgae can remove significant quantities of carbon, nitrogen, and phosphorus from wastewaters. The biochemical pathways responsible for the uptake and assimilation of these elements are obviously interrelated, making it difficult to distinguish pathways which are specific for any one element.[1] Thus, we will examine these pathways separately, followed by discussion of the pathways responsible for trace metal/algal interactions and for the uptake or adsorption of xenobiotic matter (e.g., pesticides and other man-made organic compounds) which may be found in agricultural wastewater.

II. REGULATION OF INORGANIC CARBON UPTAKE

Our knowledge of the different regulatory pathways of inorganic carbon fixation is still limited because of the diversity in phytoplankton communities and uncertainties as to the biochemical pathways involved. Probable pathways have been proposed in reviews by Appleby et al.,[2] Morris,[3] and Kremer[4] and are summarized in Figure 1.

Carbon fixation could proceed on one hand via the activity of ribulose-1,5-bisphosphate carboxylase (RuBPCase) and the Calvin-Benson cycle and on the other hand via β-carboxylation. In algae, the type of β-carboxylation differs from those described in C_4 dicarboxylic acid-producing pathways (C_4-type plants) and crassulacean acid metabolism (CAM)-type species. Indeed, although algae produce oxaloacetate after CO_2 fixation, there is no evidence that this compound is further decarboxylated and the CO_2 fed into the Calvin-Benson cycle. In fact, in the presence of light, phosphoenolpyruvate originates from 3-phosphoglycerate and from the Calvin-Benson cycle.

Three CO_2-fixing enzymes have been described in phytoplankton: phosphoenolpyruvate carboxylase (PEPCase), phosphoenolpyruvate carboxykinase (PEPCKase), and pyruvate carboxylase (PCase), which catalyze, respectively, the following reactions:

$$\text{Phosphoenolpyruvate} + \text{HCO}_3^- \xrightarrow{\text{Mg}^{2+}} \text{Oxaloacetate} + \text{P (PEPCase)}$$

$$\text{Phosphoenolpyruvate} + \text{HCO}_3^- + \text{ADP} \xrightarrow{\text{Mn}^{2+}} \text{Oxaloacetate} + \text{ATP (PEPCKase)}$$

$$\text{Pyruvate} + \text{HCO}_3^- + \text{ATP} \rightarrow \text{Oxaloacetate} + \text{ADP} + \text{P}_i \text{ (PCase)}$$

Each species displays only one enzyme system capable of catalyzing β-oxidations (Table 1). Thus, those species with PEPCKase activity are probably better adapted to low light intensities because the energy is kept in the form of ATP.[2]

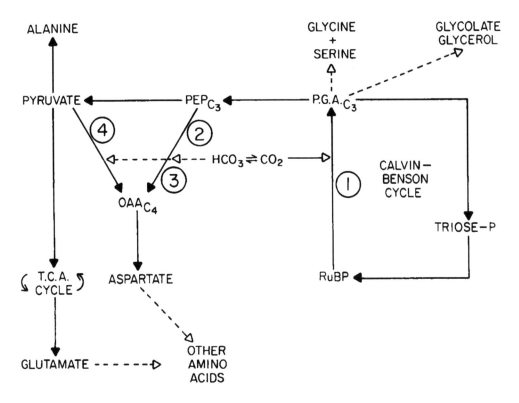

FIGURE 1. Main pathways of carbon dioxide fixation by phytoplankton cells. TCA = tricarboxylic acid, OAA = oxalocetate, PEP = phosphoenolpyruvate, PGA = 3-phosphoglycerate, Triose-P = triose-phosphate, RuBP = ribulose 1,5-biphosphate; (1) ribulose 1,5-biphosphate carboxylase; (2) phosphoenolpyruvate carboxylase; (3) phosphoenolpyruvate carboxykinase; (4) pyruvate carboxylase. (From Appleby, G., Holdsworth, E. S., and Wadman, H., *J. Phycol.*, 16, 290, 1980. With permission.)

The occurrence of a double system to fix CO_2 via the Calvin-Benson cycle and β-carboxylation is widely admitted, but the dominance of one pathway over the other is subject to controversy. Many studies suggest that RuBPCase is the key enzymatic system,[2,5-8] while others suggest that β-carboxylations are probably responsible for most of the CO_2 fixation.[9,10] As indicated by Morris,[1] phytoplankton species probably display a spectrum of metabolic types which can be described by the RuBPCase/PEPCase ratio. This ratio also describes the relationship between the photosynthetic pathways and the pattern of carbon accumulation by the cells.[11]

III. NITROGEN UPTAKE PATHWAYS

The nitrogen uptake pathway used by a phytoplankton species is dependent on the form of nitrogen being taken up. Gaseous forms can be fixed by the cyanobacteria. Biochemically, nitrogen fixation takes place through reduction of N_2 to ammonium. Syrett[12] indicated that six electrons are required for each molecule of N_2 reduced; the electrons are generally derived from reduced ferredoxin.

Inorganic nitrogen is present in fresh or marine waters in the form of nitrate (NO_3), nitrite (NO_2), or ammonium (NH_4). The ability to use any of these forms appears to be general among algae; however, several flagellates are unable to use NO_3 or NO_2 for growth.[13,14] When taken up as NO_3 or NO_2, nitrogen must be reduced before it can be incorporated into organic molecules (Figure 2). This reduction occurs via the activity of two enzymes: nitrate reductase and nitrite reductase.

TABLE 1
**Distribution of β-Carboxylation Enzymes in the
Various Groups of Phytoplankton**

Species	β-Carboxylation enzyme			
	PEPC	PEPCK	PC	Ref.
Bacillariophyceae				
Chaetoceros calcitrans	•[a]			2
Coscinodiscus sp.	•			170
Cylindrotheca closterium			•	7
			•	2
	•			10
Phaeodactylum tricornutum	•			9
			•	7
	•			10
	•			170
			•	2
Skeletonema costatum	•			9
Thalassiosira pseudonana		•		7
	•			170
			•	2
Bangiophyceae				
Porphyridium cruentum	•[a]			2
Chlorophyceae				
Dunaliella tertiolecta	•			9
	•			169
	•			170
	•[a]			2
Coccolithophorideae				
Coccolithus pelagicus	•			170
Cryptophyceae				
Chroomonas salina	•			170
Cyanophyceae				
Anabaena cylindrica	•[a]			2
Oscillatoria thiebautii	•			171
Dinophyceae				
Amphidininum carterae			•	2
Gymnodinium sp.			•	2
Gonyaulax tamarensis	•			9
Prymnesiophyceae				
Pavlova lutheri			•	2

[a]   Stimulated by Mn^{2+}.

There are two types of nitrate reductase known in algae. The first one, found in eukaryotic algae, resembles that form found in higher plants in many respects and catalyzes the reaction

$$NO_3^- + NAD(P)H + H^+ \rightarrow NO_2^- + NAD(P)^+ + H_2O$$

The enzyme has been purified and characterized from *Chlorella*[15,16] and from the diatom *Thalassiosira*.[17] The second type of nitrate reductase is found in prokaryotic cells. This type does not use pyridine nucleotide as the electron donor, but instead uses reduced ferredoxin (Fd_{red}):

$$NO_3^- + 2\,Fd_{red} + 2\,H^+ \rightarrow NO_2^- + 2\,Fd_{ox} + H_2O$$

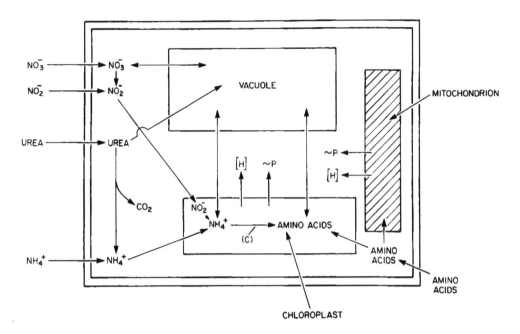

FIGURE 2. Main features of nitrogen assimilation in a eukaryotic cell. (From Syrett, P. J., *Can. Bull. Fish. Aquat. Sci.*, 210, 182, 1981. With permission.)

Nitrite reductase, which reduces nitrite to ammonia without the release of any free intermediates, is not as well known as nitrate reductase. The reaction requires reduced ferredoxin as the electron donor:

$$NO_2^- + 6\,Fd_{red} + 2\,H^+ \rightarrow NH_4^+ + 6\,Fd_{ox} + 2\,H_2O$$

In algae, the enzyme has been studied using *Anabaena*,[18] *Dunaliella*,[19] *Ditylum*,[20] *Chlorella*,[21] and *Skeletonema*.[22]

Direct assimilation of ammonium by algae has been studied extensively (see review by Syrett[12]) and leads to glutamine and glutamic acid through the action of either glutamic dehydrogenase (GDH) or the combined activities of glutamineoxoglutarate aminotransferase (GOGAT, also called glutamate synthase) and glutamine synthetase (GS). Studies have indicated that both GDH and GS/GOGAT can act as the primary NH_4^+ assimilatory pathway in a variety of species, and as yet there is no clear consensus as to which is more important in the algae.[12] In some cyanobacteria which lack GDH activity, alanine is formed from the reaction of NH_4^+ with pyruvic acid via the activity of the enzyme alanine dehydrogenase.[23]

The uptake and assimilation of inorganic N compounds by algae is affected by several environmental parameters, principally ammonium concentration and light. Inactivation of nitrate uptake by the presence of ammonium has been observed in most exponential growth-phase batch cultures studied.[24] Active nitrate reductase is not formed in the presence of ammonium, and even if it is present, the addition of ammonium can lead to a quick inhibition of nitrate utilization (see review by Syrett[12]). Light seems to influence the uptake and assimilation of N by stimulating the uptake and reduction of nitrate, nitrite, and hydroxylamine. This could be attributed to a direct photoreduction of the N compounds,[25] to the supply of energy via photophosphorylation,[26] or by an indirect effect of photosynthesis providing C skeletons to accept reduced N.[27] In most cases, nitrate reduction seems more sensitive to light activation.

IV. PHOSPHORUS UPTAKE PATHWAYS

Although the concentration of organic phosphates in natural water often exceeds that of inorganic phosphate, the major form in which microalgal cells assimilate phosphorus is as inorganic phosphate ($H_2PO_4^-$ or HPO_4^{2-}). Organic phosphate may be used as a primary source of phosphorus, but must be hydrolyzed by extracellular enzymes such as phosphoesterases[28] or phosphatases.[29]

Phosphate uptake by algae is an energy-dependent reaction and is generally stimulated by light.[30,31] The maximum rate of uptake is also dependent on the availability of Na^+, K^+, or Mg^+ ions. Whether these ions are required for the uptake process itself or for subsequent polyphosphate synthesis is still unclear.[32,33]

Phosphate in the cell is channeled into polyphosphate and various organic compounds: nucleotides such as ATP, nucleic acids, phospholipids, and sugar phosphates. While polyphosphate appears to be the principal form of phosphorus storage, some excess phosphate is stored in ionic form in vacuoles.[31] It is interesting to note that only when external phosphate concentrations become limiting is polyphosphate degraded to fulfill the metabolic needs of the cell.[34,35]

V. IMPORTANCE OF THE N:P ATOMIC RATIO

Nitrogen and phosphorus are the major nutrients for microalgal growth and, depending upon their relative availability, one or the other may become the limiting nutrient.[36,37] The N:P ratio (by atoms), R_c,[37] or optimum nutrient ratio[38] is the ratio at which a transition from one limiting nutrient to the other occurs. It is important to express this ratio in terms of atoms[37,38] to avoid any difficulties which could arise when the ratio is expressed in terms of mass (see Kunikane et al.[39]). In nutrient media with N:P ratios above R_c, P limitation will occur, and in those with ratios below R_c, the microalgae will experience N limitation. It is also important to realize that R_c varies as a function of specific growth rate,[37,40] with species,[37,38] and with both light and temperature,[38] although in the last case the variability may be secondary through light and temperature influence on growth rate. Thus, in mixed species cultures, such as occurs in many outdoor microalgal wastewater treatment facilities or in natural waters, the N:P ratio of the medium will have a direct influence on the relative abundance of species present and may even allow a single species to outcompete the others for the available resources (see Maestrini and Bonin[41]).

VI. DISSOLVED ORGANIC MATTER

In general, agricultural drainage waters contain lower concentrations of dissolved organic matter (DOM) than urban waste effluents. Nevertheless, the DOM present, which includes humic substances and similar compounds, may affect microalgal growth and the ability of the algae to remove some wastewater components, especially trace metals and phosphate. Also, microalgae may be able to remove some portion of the DOM, such as pesticides and other xenobiotics (see below).

DOM has been shown to affect microalgal metabolism in a variety of ways.[41-43] In some reports there is evidence for stimulation of algal metabolic processes,[44-46] and in others there is evidence for suppression.[47] It has frequently been observed that DOM can behave as a chelating agent that either binds and neutralizes or renders toxic metals less available.[48] Organic substances may also bind some essential oligoelements. It has also been reported that DOM can be used directly as a source of growth factors or of essential elements[45] and that they can increase cell permeability and, hence, allow greater nutrient uptake.[44] It has been shown for *Selenastrum capricornutum* that some fractions of DOM (especially the 10^4

to 10^5 mol wt range fraction) do impair the bioavailability of certain essential nutrients (other than N, P, and C) during the exponential phase of growth.[49] Moreover, these same authors have shown that the 10^3 to $>10^5$ mol wt fraction of the DOM is able to reduce phosphate utilization during exponential growth.

Interactions between DOM, nutrients, and trace metals are complex, and the literature is filled with seemingly contradictory results which report that DOM impacts microalgae positively by increasing the bioavailability of essential macro and trace elements[50,51] or negatively by decreasing bioavailability.[52] It is quite clear that one cannot reliably predict the effects of DOM on algal growth in a given type of wastewater. Direct case-by-case experimental evidence for a given effluent has to be gained before any specific conclusions can be drawn.

VII. TRACE METALS

A. PHYSIOLOGICAL REQUIREMENTS

Many algal biochemical pathways require trace metals such as Fe, Mn, Co, Cu, Zn, and Mo as cofactors in enzyme-mediated reactions. In addition, metals like Mg, Fe, and Cu are also components of vital and active compounds such as the chlorophylls (Mg), cytochromes and ferredoxin (Fe), plastocyanin (Cu), and a suite of metalloproteins and metalloflavoproteins.[53,54] Metals play a dualistic role in algal physiology — each metal is required over a fairly narrow concentration range, while concentrations either above or below this range can be toxic or limiting to growth.[55] The effects of metals on algal growth and physiology are complex due to a number of metal-metal or metal-algal interactions which will be discussed below. Another serious stumbling block to the understanding of trace metal effects on microalgae has been our inability, until recently, to accurately measure environmental concentrations of trace metals. In this section, we will examine the trace metal-algal interactions which will either effect or enable microalgae to remove trace metals occurring in agricultural wastewaters.

A comprehensive review of trace metals in the environment is beyond the scope of this section. Bruland[56] presents the most comprehensive review to date of trace metals in seawater. In addition, Huntsman and Sunda[55] and references cited therein discuss the role trace metals play in microalgal growth. It is important to understand the complex chemistry of trace metals if one is interested in the study of trace metal-algal interactions. The above references will be especially useful in this respect.

B. MEASUREMENT TECHNIQUES

The development of clean sampling and analytical techniques[57,58] has enabled researchers for the first time to derive accurate measurements of the environmental concentrations of trace metals free of contamination. Concentrations of trace metals are now believed to be in the nanomolar range, whereas prior to the development of the more sensitive analytical techniques, concentrations in seawater were thought to be one to several orders of magnitude greater. These techniques represent a major step in the understanding of trace metal-microalgal interactions because we can now collect contamination-free samples and examine the effects of additions of trace metals in the appropriate concentration ranges.[59-61]

Another step forward in the understanding of trace metal-microalgal interactions is the realization that most frequently it is the concentration of the free metal ion that controls bioavailability.[55,62-65] This requires that one must determine the degree of complexation for each of the chemical species of a particular metal in addition to the concentrations of the free ion.[66-69] Trace metal speciation is difficult to assess because of the relative insensitivity of the techniques commonly applied.[70,71] Techniques such as anodic stripping voltometry, fixed potential amperometry, bioassays, and ion-specific electrodes also are very difficult to calibrate in that the degree of speciation affects calibration relationships.[64,65,70,72-75] Because of problems in calibration for all of the above techniques, Sunda and Hanson[65] suggest the combined use of C_{18} SEP-PAK cartridges (Millipore Corp., Bedford, MA) and ligand exchange based on internal standard calibration, which appears to have the required specificities and sensitivities to determine free Cu ion concentrations.[65]

C. COMPLEXATION

A variety of natural substances have been shown to form organic complexes with trace metals. Other chelators, such as humic and fulvic acids, are generally considered to have the primary role in the complexation of metals in freshwater and seawater.[76-81] It is also becoming clear that organometal complexes are generally not available for uptake and utilization by organisms; however, in the case of Hg, for example, organometal complexes are frequently more toxic than the free ion.[55] A notable exception is vitamin B_{12}, which contains Co and Fe hydroxamates that are used by cyanobacteria.[82-86] In addition, complexation with organic chelators may promote the reduction of Fe and Mn to oxidation states which are more available to organisms.[55] Redox enzymes associated with the plasmalemma in marine phytoplankton have also been shown to reduce Fe and Cu to more favorable oxidation states.[87] Thus, reactions with organic molecules, either at the plasmalemma or in the surrounding environment, can act to either complex with trace metals, thereby decreasing toxicity and making the metals biologically unavailable, or, in some cases, reduce the metal to an oxidation state which is more available.

D. COMPETITIVE INTERACTIONS

It has become apparent that trace metals interact competitively or antagonistically with other metals, complicating interpretation of results in the study of trace metal effects on algal growth.[55,88-90] It has been known for some time that Co and Fe compete for binding sites in microalgae.[91-95] Another pair of trace metals which have received much attention are Cd and Fe.[88,96] In the above metal-metal interactions, it appears that one metal preferentially binds to the cell uptake sites, thereby reducing the availability of the second metal to the cell. Thus, the cell will experience the effects of deficiency of the second metal.[55]

There appears to be a link between nutrient dynamics and trace metal availability.[55] For example, the metals Fe and Cu have been shown to play roles in phytoplankton production and the uptake dynamics of both nitrogen and silica.[97-108] It is known that the enzyme nitrate reductase contains Fe[98] and that Fe deficiency in *Scenedesmus* sp. results in decreased nitrate reductase activity.[107] Rueter and Ades[106] present evidence that nitrate uptake is a function of both light and Fe concentration for the chlorophyte *Scenedesmus quadricauda;* Fe-limited cultures exhibit decreased rates of nitrate uptake in low light, while ammonium uptake is unaffected.[106] It is also apparent that the pathways of photosynthetic carbon fixation into major end products are affected by Fe limitation. Growth limitation by Fe results in the increased synthesis of protein coincident with decreased nitrate uptake.[99,106] The results of the above studies all indicate that the carbon and nitrogen assimilatory pathways are affected by both trace metal deficiency and toxicity.

E. ADSORPTION BY ALGAE OR ALGAL EXTRACTS

The cell wall polysaccharides of several species of algae affect ion-exchange reactions with polyvalent metals. Fucoidin from *Ascophyllum nodosum* has been found to bind Pb more strongly than the other divalent cations in the series tested (e.g., Pb, Ba, Cd, Sr, Cu, Fe, Co, Zn, Mg, Mn, Cr, Ni, Hg, Ca).[109] κ-Carrageenan from *Eucheuma striatum* and *E. spinosum* strongly binds Pb and Cd.[110] Alginic acid from brown algae has high affinity for Sr. Most studies with alginates assess their inhibitory effect on intestinal absorption of radioactive Sr, the most inhibitory species being *Macrocystis pyrifera, Egregia menziesii,*

Nereocystis luetkana, Alaria marginata, and *Laminaria digitata.*[111] In nature, concentration factors greater than 1000 to 1 from seawater for trace metals such as Cu, Zn, Pb, and Mn have been found in *Fucus vesiculosus* and *Laminaria digitata.*[112] Brown algae such as those that concentrate trace metals from seawater have been examined as environmental monitors.[113]

Cell surfaces of microalgae show strong adsorption of metal ions, including Pb, Cu, Ni, Zn, Cd, Ag, Hg, and U.[114] Biomass of *Chlorella vulgaris* has been used for recovery of Au, Ag, Hg, and U ions.[115-117] Lyophilized preparations of whole algae have been applied to industrial waste treatment. Whole, dried cells of *Chlorella vulgaris* complexed to silica gel act as ionic exchangers for Cu; high concentrations of Cu, particularly from electroplating wastewater, bind strongly to the algal cells and can thereby be removed.[117]

F. DETOXIFICATION

Virtually all living organisms have evolved mechanisms to counter the toxic effects of exposure to heavy metals in the environment.[118-120] Animals, plants, and fungi all synthesize cysteine-rich metal-binding proteins which selectively chelate heavy metal ions such as Cu, Zn, and Cd.[121] This group of proteins is further characterized by its inducibility in the presence of these ions.[122] The genes coding for metallothioneins appear to be members of a larger gene family containing sequences and coding proteins with specificities for particular metal ions. Although these characteristics generally appear to be highly conserved from fungi through the higher vertebrates, there are marked differences in nucleotide sequences, amino acid sequences, and protein structure across these taxa.

In plants — including the algae — there is a class of heavy-metal-binding proteins which shares the above characteristics, but because of some notable differences from the metallothioneins (which are not detected in plants), the proteins which have been observed to chelate heavy metal ions in plants have recently been labeled "phytochelatins".[123] The characteristic which appears to differentiate the metallothioneins from the phytochelatins is the involvement of glutathione (Glu[-CysGly]) and phytochelatin biosynthesis. Indeed, phytochelatin from a taxonomically broad spectrum of plants is a small peptide (2000 to 4000 mol wt) of 5 to 17 amino acids consisting of 2 to 11 repetitive γ-glutamylcysteines with a glycine at the carboxyl terminus. Grill et al.[124] conclude that synthesis of this heavy-metal-binding peptide is achieved by the sequential addition of γ-glutamylcysteine residues to glutathione, as opposed to the metallothioneins, which are direct gene products.

Functionally, phytochelatins and metallothioneins appear to be analogous. Although synthesis of phytochelatin is most actively induced by Cd, Zn, Ag, Pb, Sb, Cu, and Hg ions, the cations of Au, Bi, Sn, and Ni also induce the biosynthesis of phytochelatin, as do the anions AsO_4^{3-} and SeO_3^{2-}. Intracellularly, Cd, Zn, Pb, and Hg ions are usually found only when chelated with phytochelatin.

In order for microalgae to be useful in the treatment of agricultural wastewaters, one must consider how these organisms adapt to trace metal stress. Some organisms, such as the diatom *Phaeodactylum tricornutum,* release extracellular products which complex with metals such as Pb.[125] Other species, such as *Thalassiosira pseudonana, Skeletonema costatum,* and *Cricosphaera elongata,* release products which can complex with Cu.[126-128] It has also been shown in a number of cases (reviewed by Huntsman and Sunda[55]) that media which contain extracellular release products of microalgae, such as those above, have the capacity to complex with metals, even when the media are being used to grow another organism. Another adaptive mechanism which microalgae have developed to detoxify metals involves plasmalemma enzymes that reduce trace metals.[87] Huntsman and Sunda[55] discuss several other adaptations to trace metals found in phytoplankton, including changes in cell wall permeability to metals and compartmentalization of trace metals into inert intracellular sites. Thus, microalgae are useful in detoxifying trace metals contained in agricultural wastewater, primarily through their capacity to generate and release extracellular materials which bind with the metals. Once bound, these metals are generally no longer available to the microalgae.

VIII. REMOVAL OF XENOBIOTICS

There are essentially three mechanisms for removing xenobiotic (man-made or foreign organic) compounds from a liquid medium through the use of algae. First, algae present in the medium can metabolize or degrade xenobiotics. The second and third processes are adsorption and absorption of the compounds onto or into the algae. In all cases, these algae must be removed from the medium in order to remove the target compounds, unless the metabolism can proceed to the point of generating nontoxic secondary metabolites. In nature, a combination of the above processes occurs. Material trapped on algae that falls to the sediment can then be (further) degraded by the bacterial community. Three studies address the role of algal assemblages in the uptake of hydrocarbons from the environment.[129-131] Several general reviews on the interaction of algae with pesticides and herbicides have been compiled.[132-135]

Relatively little published information exists on the algal degradation of pesticides, herbicides, and related compounds. Presented in Table 2 is a list of the algal strains which have demonstrated the ability to metabolize one or more such compounds. Organisms which were studied, but which did not demonstrate such abilities, were not included. Only the primary or final degradation products were listed. These studies were carried out in the laboratory using axenic batch cultures with continuous light. When several culture conditions were provided, results from those most closely resembling outdoor, mass culture conditions were selected. As can be seen from the limited number of studies, metabolic degradation proceeds very slowly in algae.

Virtually no work has been published regarding the enzymatic processes involved in the metabolism of xenobiotics by microalgae, although bacterial processes are thought to apply in most cases. Reviews on microbial degradation pathways for pesticides and herbicides are given in several texts.[157,158] Narro et al.[159] listed the types of reactions involved in the transformation of aromatic compounds by algae and speculated that these are enzymatic processes. Matsumura and Esaac[153] found that a heat-stable photochemical reaction was involved in the degradation of insecticides, which in some cases was enhanced by the addition of a flavin cofactor (flavin mononucleotide). Other degradation processes were reported to be light insensitive.[144] Metabolism of various compounds to CO_2 was described in a number of reports;[139,141,147] the recovery to this level ranged from 0.1 to 3.0%. In fact, some of the metabolic intermediates were shown to be more toxic to the alga than the initial material.[152]

Studies on the ability of algae to bioconcentrate pesticides, herbicides, and related compounds are presented in Table 3. The issue of whether these xenobiotics are absorbed or adsorbed has been very difficult to resolve definitively[162] and has not been addressed in this table. Compounding the difficulty in measuring adsorbed/absorbed material is the fact that desorption of xenobiotics is enhanced when cells die. Because many of these compounds are toxic to algae, the cell turnover rate is increased along with the related desorption processes. Veber et al.[163] and Rice and Sikka[145] demonstrated very rapid (within minutes) uptake of xenobiotics from the medium, but then a slow release of the compound back into the medium, in systems where toxic effects were observed. Gross xenobiotic compound uptake was not related specifically to cell growth in the above case, but to culture biomass.

Uptake of compounds has been expressed in terms of relative concentration of the cells and the xenobiotic in the medium, the percent removal of the compound from the medium, and the unit weight of compound per unit weight of alga or chemical component of the alga. In order to adequately describe the potential rate of removal, it is necessary to relate the

TABLE 2
Genera of Algae Reported to Metabolize Pesticides/Herbicides and Related Xenobiotics

Division/genus	Substrate and major product	Rates measured	Ref.
Chlorophyta			
Ankistrodesmus amalloides	DDT to DDE	3.5% in 30 d	136
A. braunii	Simazine detoxication		137
A. falcatus	Tri-*n*-butyltin to di-*n*-butytin	50% in 4 weeks	138
Chlamydomonas ulvalensis	Catechol	2.5% as CO_2 in 48 h	139
C. angulosa	Naphthalene to 1-naphthol	1.3% per day	140
Chlorella eutrophica	Naphthalene to 1-naphthol	1.5% per day	140
C. sorokiniana	Naphthalene to 1-naphthol	1.2% per day	140
C. pyrenoidosa	Catechol	1.8% as CO_2 in 48 h	139
Chlorosarcina sp.	Simazine detoxication		137
Dunaliella tertiolecta	Cleave phloroglucinol ring	2.5% as CO_2 in 16 h	141
	DDT to DDE	7.4% in 9 d	142
	Napththalene to 1-naphthol	1.0% per day	140
Dunaliella sp.	DDT to TDE, DDOH, and DDE	18.4, 6.9, and 5.9% in 30 d	143
	Dieldrin to photodieldrin	8.5% in 30 d	143
	Aldrin to dieldrin	23.2% in 30 d	143
Platymonas (Tetraselmis) sp.	Cleave phloroglucinol ring	1.9% as CO_2 in 16 h	141
Scenedesmus basiliensis	Catechol	5.1% as CO_2 in 48 h	139
	Phenol	0.5% as CO_2 in 48 h	139
S. quadricauda	2,4-D to 3-OH-2,4-D	41% in 24 h	144
Tetraselmis chuii	DDT to DDE	11.5% in 24 d	145
Ulva fasciata	Naphthalene to 1-naphthol	1.2% per day	140
Euglenophyta			
Euglena gracilis	Catechol	0.6% as CO_2 in 48 h	139
	Phenol	0.3% as CO_2 in 48 h	139
Phaeophyta			
Petalonia fascia	Naphthalene to 1-naphthol	1.9% per day	140
Chrysophyta			
Amphora sp. AMP-1	Naphthalene to 1-naphthol	1.2% per day	140
Cyclotella nana	DDT to DDE	<3.0% in 14 d	142
	DDT to DDE	1.2% in 24 d	142
Cylindrotheca closterium	DDT to DDE	8.8% in 21 d	146
C. fusiformis	Cleave phenylalanine ring	0.37% as CO_2 in 14 d	147
Cylindrotheca sp. N-1	Naphthalene to 1-naphthol	1.4% per day	140
	Naphthalene to 1-naphthol	0.7% per day	148
Isochrysis galbana	Cleave phenylalanine ring	0.05% as CO_2 in 14 d	147
	DDT to DDE	0.7% in 24 d	145
Navicula incerta	Cleave phenylalanine ring	1.5% as CO_2 in 14 d	147
Navicula sp. K1A	Naphthalene to 1-naphthol	0.8% per day	148
Nitzschia angularis	Cleave phenylalanine ring	0.6% as CO_2 in 14 d	147
Nitzschia sp. K8A	Naphthalene to 1-naphthol	0.8% per day	148
Nitzschia sp.	DDT to DDE	0.4% in 14 d	149
Olisthodiscus luteus	DDT to DDE	0.9% in 24 d	145
Phaeodactylum tricornutum	Cleave phloroglucinol ring	1.9% as CO_2 in 16 h	141
	Cleave phenylalanine ring	0.1% as CO_2 in 14 d	147
Prymnesium parvum	Cleave phenylalanine ring	0.1% as CO_2 in 14 d	147
Skeletonema costatum	Cleave phloroglucinol ring	0.2% as CO_2 in 16 h	141
	DDT to DDE	4.5% in 14 d	142
	DDT to DDE	3.3% in 10 d	145
Synedra sp. 4D	Naphthalene to 1-naphthol	1.4% per day	148
Syracosphaera carteri	Cleave phloroglucinol ring	3.0% as CO_2 in 16 h	141
Thalassiosira fluviatilis	DDT to DDE	3.9% in 11 d	142
Pyrrophyta			
Amphidinium carteri	Cleave phloroglucinol ring	5.5% as CO_2 in 16 h	141
	Cleave phenylalanine ring	0.1% as CO_2 in 14 d	147
	DDT to DDE	2.8% in 24 d	145

TABLE 2 (continued)
Genera of Algae Reported to Metabolize Pesticides/Herbicides and Related Xenobiotics

Division/genus	Substrate and major product	Rates measured	Ref.
Rhodophyta			
Porphyridium sp.	Cleave phloroglucinol ring	5.0% as CO_2 in 16 h	141
P. cruentum	Naphthalene to 1-naphthol	1.2% per day	140
Cryptophyta			
Rhodomonas lens	Cleave phenylalanine ring	0.9% as CO_2 in 14 d	147
Cyanophyta			
Agmenellum quadruplica-	Cleave phenylalanine ring	0.1% as CO_2 in 14 d	147
tum	Naphthalene to 1-naphthol	1.4% per day	150
	Naphthalene to 1-naphthol	1.4% per day	140
	Aniline to formanilide, etc.	1.3% per day	151
	1- or 2-methylnaphthalene to 1- or 2-hydroxymethylnaphthol	1.9% per day of 1-hydroxymethylnaphthol; 3.2% per day of 2-hydroxymethylnaphthol	152
Anabaena cylindrica	Catechol	0.5% as CO_2 in 48 h	139
Anabaena sp. CA	Naphthalene to 1-naphthol	2.0% per day	140
	1- or 2-methylnaphthalene to 1- or 2-hydroxymethylnaphthol	0.9% per day of 1-hydroxymethylnaphthol; 0.7% per day of 2-hydroxymethylnaphthol	152
Anabaena sp. 1F	Naphthalene to 1-naphthol	1.5% per day	140
Anacystis nidulans TX-70	DDT to DDE; lindane to BTC	Not quantified	153
	Dieldrin to photodieldrin	Not quantified	153
	Mexacarbate to water-soluble component (WSC)	0.3 nmol per mg protein per 2 min	153
	Parathion to WSC	Degradation by protein extracts	153
	Toxaphene to WSC	Degradation by protein extracts	153
Aphanocapsa sp. 6714	Naphthalene to 1-naphthol	0.5% per day	140
Coccochloris elabens	Naphthalene to 1-naphthol	2.0% per day	140
Microcystis aeruginosa	Mexacarbate to WSC	Degradation by protein extracts	153
Microcoleus chthonoplastes	Naphthalene to 1-naphthol	0.2% per day	140
Nostoc sp. MAC	Naphthalene to 1-naphthol	0.1% per day	140
Oscillatoria sp. JCM	Naphthalene to 1-naphthol	2.4% per day	154
	Naphthalene to 1-naphthol	2.4% per day	140
	Biphenyl to 4-hydroxybiphenol	2.9% per day	155
	Aniline to formanilide, etc.		151
	1- or 2-methylnaphthalene to 1- or 2-hydroxymethylnaphthol	1.8% per day of 1-hydroxymethylnaphthol; 1.4% per day of 2-hydroxymethylnaphthol	152
Oscillatoria sp. MEV	Naphthalene to 1-naphthol	0.3% per day	140
Phormidium foveolarum	Catechol	0.4% as CO_2 in 48 h	139
P. fragile	Methomyl	48.6% degradation	156

Note: Time factor reflects duration of experiment rather than metabolic rate.

concentration of the xenobiotic material removed to the algal biomass. Cox[168] related DDT uptake to cell carbon in three algal species. Matsumura and Esaac[153] described pesticide degradation in terms of nanomoles original product metabolized per milligram flavoprotein per unit of time. Only Sodergren[162] and Rice and Sikka[145] described pesticide uptake per unit dry weight over the time period when maximum uptake occurs. They demonstrated that the uptake essentially reached a plateau after 1 h of exposure of the cells to DDT. Removal of xenobiotics from water into algae has been reported to approach 100% (e.g., for DDT,[142] pesticides,[164] and atrazine[163]). While the algae vary in their ability to accumulate specific compounds, the bioconcentration process is clearly very efficient. Thus, we can conclude that wastewater cleanup based on the removal of xenobiotics by algal systems is feasible if the algae can be economically separated from the runoff water and disposed of. These processes are discussed elsewhere in this volume.

TABLE 3
Bioconcentration of Xenobiotics by Algae

Division/genus	Compound concentrated	Rate	Ref.
Chlorophyta			
Ankistrodesmus amalloides	DDT (0.7 ppb)	61,600 X^a in 3 h	136
	Dieldrin (0.7 ppb)	32,000 X^a in 3 h	136
	Photodieldrin (0.7 ppb)	4,700 X^a in 3 h	136
A. falcatus	Tri-n-butyltin (20 μg/l)	30,000 X^a	138
Chlamydomonas reinhardtii	2,4-D (0.1 ppm; pH = 7)	4.0% in 6 h[b]	144
Chlamydomonas sp.	Mirex (up to 50 pptr)[c]	3,200 X^a	160
Chlorella pyrenoidosa	2,4-D (0.1 ppm; pH = 7)	6% in 6 h[b]	144
	Lindane (180 ng/l)	39% in continuous culture[b]	161
	DDE (98 ng/l)	82% in continuous culture[b]	161
	PCB (3,700 ng/l)	88% in continuous culture[b]	161
	DDT (0.6 μg/l)	0.32 μg per mg dry weight	162
C. vulgaris	Atrazine (0.25 to 15 ppm)	95 to 98%[b]	163
Chlorella (11 spp.)	2,4-D (0.01 ppm)	71.6% ± 12.9 in 14 d[b]	164
	Diazinon (1 ppm)	13.5% ± 9.5 in 14 d[b]	164
	Methylchlor (0.01 ppm)	40.0% ± 15.3 in 14 d[b]	164
Chlorococcum sp.	Mirex (up to 50 pptr)[c]	7,300 X^a	160
Dunaliella tertiolecta	DDT (80 ppb)	94.6% in 3 d[b]	142
	Mirex (up to 50 pptr)[c]	4,100 X^a	160
Scenedesmus obliquus	DDT (1 ppm)	626 X^b in 7 d	165
	Parathion (1 ppm)	72 X^b in 7 d	165
S. quadricauda	Chlordane (0.1 to 100 μg/l)	All about 10,000 X^a	166
	2,4-D (0.1 ppm; pH = 7)	2% in 6 h[b]	144
Scenedesmus (2 spp.)	2,4-D (0.01 ppm)	75% in 14 d[b]	164
	Diazinon (1 ppm)	18.5% ± 2.1 in 14 d[b]	164
	Methylchlor (0.01 ppm)	60.5% ± 3.5 in 14 d[b]	164
Tetraselmis chuii	DDT (0.7 ppb)	4.0 ng DDT per mg dry weight per 2 h	145
Euglenophyta			
Euglena gracilis	DDT (1 ppm)	99 X^b in 7 d	165
	Parathion (1 ppm)	62 X^b in 7 d	165
Chrysophyta			
Coccolithus huxleyi	DDT (80 ppb)	85.1% in 4 d[b]	142
Cylindrotheca closterium	DDT	190 X^a	146
	PCB (0.01 ppm)	1,100 X^a	167
Cyclotella nana	DDT (80 ppb)	93.3% in 3 d[b]	142
	DDT (0.7 ppb)	14.5 ng DDT per mg dry weight per 2 h	145
Isochrysis galbana	DDT (0.7 ppb)	12.5 ng DDT per mg dry weight per 2 h	145
Nitzschia (3 spp.)	2,4-D (0.01 ppm)	82.4% ± 4.0 in 14 d[b]	165
	Diazinon (1 ppm)	21.0% ± 5.3 in 14 d[b]	165
	Methylchlor (0.01 ppm)	56.7% ± 14.0 in 14 d[b]	165
Olisthodiscus luteus	DDT (0.7 ppb)	12.0 ng DDT per mg dry weight per 2 h	145
Skeletonema costatum	DDT (80 ppb)	95.7% in 17 d[b]	142
	DDT (0.7 ppb)	18.4 ng DDT per mg dry weight per 2 h	145
Syracosphaera carteri	DDT (2.3 ppb)	25,000 X^a	168
Thalassiosira fluviatilis	DDT (3.0 ppb)	25,000 X^a	168
	DDT (80 ppb)	99.5% in 8 d[b]	142
Thalassiosira pseudonana	Mirex (up to 50 pptr)[c]	5,000 X^a	160
Pyrrophyta			
Amphidinium carteri	DDT (2.7 ppb)	80,000 X^a	168
	DDT (80 ppb)	90.9% in 7 d[b]	142
	DDT (0.7 ppb)	6.6 ng DDT per mg dry weight per 2 h	145
Cyanophyta			
Anacystis nidulans	DDT (1 ppm)	849 X^a in 7 d	165
	Parathion (1 ppm)	50 X^a in 7 d	165

TABLE 3 (continued)
Bioconcentration of Xenobiotics by Algae

Note: Time factors reflect duration of experiment rather than uptake kinetics; quantity in parentheses is concentration of compound at start of experiment.

[a] Concentration of compound in cell relative to medium.
[b] Amount of original quantity of compound removed from the medium.
[c] Parts per trillion.

IX. SUMMARY

From discussion of the above topics, it is clear that microalgae have the appropriate biochemical pathways to make them potentially useful for the reclamation of agricultural wastewaters. Phytoplankton have the capability to take up significant quantities of dissolved forms of both nitrogen and phosphorus. Large-scale culture of microalgae to remove dissolved N and P can present serious engineering problems, especially in terms of circulation of the culture and harvesting techniques. Problems also arise in terms of species succession in outdoor cultures, especially when a particular species with certain uptake and growth characteristics is desired. Both of the above topics are addressed elsewhere in this volume.

Microalgae also appear to have the capability of degrading, adsorbing, or absorbing various pesticides, herbicides, and other xenobiotic organic compounds. It is still not clear whether algae can remove xenobiotics efficiently without significant deleterious effects on their growth capabilities, but the finding that microalgae can remove these sorts of agricultural wastewater contaminants is clearly important. Further studies, including the use of genetic engineering (see elsewhere in this volume for a discussion of this topic), may yield an algal species capable of efficient growth as well as significant removal of organic contaminants in drain waters.

REFERENCES

1. **Morris, I.,** Photosynthesis products, physiological state and phytoplankton growth, in *Physiological Bases of Phytoplankton Ecology*, Platt, T., Ed.; *Can. Bull. Fish. Aquat. Sci.*, 1981, 210, 83.
2. **Appleby, G., Holdsworth, E. S., and Wadman, H.,** β-Carboxylation enzymes in marine phytoplankton and isolation and purification of pyruvate carboxylase from *Amphidinium carterae*, *J. Phycol.*, 16, 290, 1980.
3. **Morris, I.,** Paths of carbon assimilation in marine phytoplankton, in *Productivity in the Sea*, Falkowski, P. G., Ed., Plenum Press, New York, 1980, 139.
4. **Kremer, B. P.,** Dark reaction of photosynthesis, in *Physiological Bases of Phytoplankton Ecology*, Platt, T., Ed.; *Can. Bull. Fish. Aquat. Sci.*, 210, 44, 1981.
5. **Kremer, B. P. and Berks, R.,** Photosynthesis and carbon metabolism in marine and freshwater diatoms, *Z. Pflanzenphysiol.*, 87, 148, 1978.
6. **Holdsworth, E. S. and Colbeck, J.,** The pattern of carbon fixation in the marine unicellular alga *Phaeodactylum tricornutum*, *Mar. Biol.*, 38, 189, 1976.
7. **Holdsworth, E. S. and Bruck, K.,** Enzymes concerned with β-carboxylation in marine phytoplankton, *Arch. Biochem. Biophys.*, 182, 87, 1977.
8. **Smith, J. C., Platt, T., and Harrison, W. G.,** Photoadaptation of carboxylating enzymes and photosynthesis during a spring bloom, *Prog. Oceanogr.*, 12, 425, 1983.
9. **Beardall, J., Mukerji, D., Glover, H. E., and Morris, I.,** The path of carbon photosynthesis by marine phytoplankton, *J. Phycol.*, 12, 409, 1976.
10. **Mukerji, D. and Morris, I.,** Photosynthetic carboxylating enzymes in *Phaeodactylum tricornutum:* assay methods and properties, *Mar. Biol.*, 36, 409, 1976.

11. **Priscu, J. C. and Goldman, C. R.**, Carboxylating enzyme activity and photosynthetic end products of photoplankton in the shallow and deep chlorophyll layer of Castle Lake, *Limnol. Oceanogr.*, 28, 1168, 1983.
12. **Syrett, P. J.**, Nitrogen metabolism of microalgae, in *Physiological Bases of Phytoplankton Ecology*, Platt, T., Ed.; *Can. Bull. Fish. Aquat. Sci.*, 210, 182, 1981.
13. **Birdsey, E. C. and Lynch, V. H.**, Utilization of nitrogen compounds by unicellular algae, *Science*, 137, 763, 1963.
14. **Casselton, P. J. and Stacey, J. L.**, Observations on the nitrogen of *Prototheca kruger*, *New Phytol.*, 68, 731, 1969.
15. **Salomonson, L. P.**, Structure of *Chlorella* nitrate reductase, in *Nitrogen Assimilation of Plants*, Hewitt, E. J. and Cutting, C. V., Eds., Academic Press, New York, 1979, 199.
16. **Salomonson, L. P., Lorimer, G. H., Hall, R. L., Borchers, R., and Bailey, J. L.**, Reduced nicotinamide adenine dinucleotide-nitrate reductase of *Chlorella vulgaris:* purification, prosthetic groups and molecular properties, *J. Biol. Chem.*, 250, 4120, 1975.
17. **Amy, N. K. and Garrett, R. H.**, Purification and characterization of the nitrate reductase of the diatom *Thalassiosira pseudonana*, *Plant. Physiol.*, 54, 629, 1974.
18. **Hattori, A. and Uesugi, I.**, Purification and properties of nitrite reductase from the blue-green alga *Anabaena cylindrica*, *Plant Cell Physiol.*, 9, 689, 1968.
19. **Grant, B. R.**, Nitrite reductase in *Dunaliella tertiolecta:* isolation and properties, *Plant Cell Physiol.*, 11, 55, 1970.
20. **Eppley, R. W. and Rogers, J. N.**, Inorganic nitrogen assimilation in *Ditylum brightwellii*, *J. Phycol.*, 6, 344, 1970.
21. **Zumpt, W. G.**, Ferredoxin: nitrite oxidoreductase from *Chlorella*. Purification and properties, *Biochim. Biophys. Acta*, 276, 363, 1978.
22. **Llama, M. J., Macarulla, J. M., and Serra, J. L.**, Characterization of the nitrite reductase activity in the diatom *Skeletonema costatum*, *Plant Sci. Lett.*, 14, 169, 1979.
23. **Neilson, A. H. and Doudoroff, M.**, Ammonia assimilation in blue-green algae, *Arch. Mikrobiol.*, 89, 15, 1973.
24. **Harvey, H. W.**, Synthesis of organic nitrogen and chlorophyll by *Nitzschia closterium*, *J. Mar. Biol. Res. Assoc. U.K.*, 31, 477, 1953.
25. **Hattori, A.**, Light induced reduction of nitrate, nitrite and hydroxylamine by a blue-green alga, *Anabaena cylindrica*, *Plant Cell Physiol.*, 3, 355, 1962.
26. **Ahmed, J. and Morris, I.**, The effect of 2,4-dinitrophenol and other uncoupling agents on the assimilation of nitrate and nitrite by *Chlorella*, *Biochim. Biophys. Acta*, 162, 32, 1968.
27. **Grant, B. R.**, The action of light on nitrate and nitrite assimilation by the marine chlorophyte *Dunaliella tertiolecta*, *J. Gen. Microbiol.*, 48, 379, 1967.
28. **Reichardt, W., Overbeck, J., and Steubing, L.**, Free dissolved enzymes in lake waters, *Nature*, 216, 1345, 1968.
29. **Strickland, J. D. H. and Solorzano, L.**, Determination of monoesterase hydrolyzable phosphate and phosphorus esterase activity in seawater, in *Some Contemporary Studies in Marine Sciences*, Barnes, H., Ed., Allen and Uwin, London, 1966, 665.
30. **Healy, F. P.**, Characteristics of phosphate deficiency in *Anabaena*, *J. Phycol.*, 9, 383, 1973.
31. **Smith, F. A.**, Active phosphate uptake by *Nitella translucens*, *Biochim. Biophys. Acta*, 126, 94, 1966.
32. **Kuhl, A.**, Phosphorus, in *Algal Physiology and Biochemistry*, Stewart, W. P. D., Ed., Blackwell Scientific, Oxford, 1974, 636.
33. **Mohleji, S. C. and Verhoff, F. H.**, Sodium and potassium ion effects on phosphorus transport in algal cells, *J. Water Pollut. Control Fed.*, 52, 110, 1980.
34. **Baker, A. L. and Schmidt, R. R.**, Polyphosphate metabolism during nuclear division in synchronously growing *Chlorella*, *Biochim. Biophys. Acta*, 82, 624, 1964.
35. **Baker, A. L. and Schmidt, R. R.**, Induced utilization of polyphosphate during nuclear division in synchronously growing *Chlorella*, *Biochim. Biophys. Acta*, 93, 180, 1964.
36. **Rhee, G.-Y.**, Effects of N:P atomic ratios and nitrate limitation on algal growth, cell composition, and nitrate uptake, *Limnol. Oceanogr.*, 23, 10, 1978.
37. **Terry, K. L.**, Nitrogen and phosphorus requirements of *Pavlova lutheri* in continuous culture, *Bot. Mar.*, 23, 757, 1980.
38. **Rhee, G.-Y. and Gotham, I. J.**, Optimum N:P ratios and coexistence of planktonic algae, *J. Phycol.*, 16, 486, 1980.
39. **Kunikane, K., Kaneko, M., and Maehara, R.**, Growth and nutrient uptake of the green alga, *Scenedesmus dimorphus*, under a wide range of nitrogen/phosphorus ratio. I. Experimental study, *Water Res.*, 18, 1299, 1984.
40. **Claesson, A.**, Variation in cell composition and utilization of N and P for growth of *Selenastrum capricornutum*, *Physiol. Plant.*, 48, 59, 1980.

41. **Maestrini, S. Y. and Bonin, D. J.**, Competition among phytoplankton based on inorganic macronutrients, in *Physiological Bases of Phytoplankton Ecology*, Platt, T., Ed.; *Can. Bull. Fish. Aquat. Sci.*, 210, 264, 1981.

42. **Bonin, D. J. and Maestrini, S. Y.**, Importance of organic nutrients for phytoplankton growth in natural environments: implications for algal species succession, in *Physiological Bases of Phytoplankton Ecology*, Platt, T., Ed.; *Can. Bull. Fish. Aquat. Sci.*, 210, 279, 1981.

43. **Bonin, D. J., Maestrini, S. Y., and Leftly, J. W.**, The role of hormones and vitamins in species succession of phytoplankton, in *Physiological Bases of Phytoplankton Ecology*, Platt, T., Ed.; *Can. Bull. Fish. Aquat. Sci.*, 210, 310, 1981.

44. **Prakash, A. and Rashid, M. A.**, Influence of humic substances on the growth of marine phytoplankton: dinoflagellates, *Limnol. Oceanogr.*, 13, 598, 1968.

45. **Sachdev, D. R. and Clesceri, N. L.**, Effects of organic fractions from a secondary effluent on *Selenastrum capricornutum* (Kutz), *J. Water Pollut. Control Fed.*, 50, 1810, 1978.

46. **Tison, D. L. and Lingg, A. J.**, Dissolved organic matter utilization and oxygen uptake in algal-bacterial microcosms, *Can. J. Microbiol.*, 25, 1315, 1979.

47. **Jackson, T. A. and Hecky, R. E.**, Depression of primary productivity by humic matter in lake and reservoir waters of the boreal forest zone, *Can. J. Fish. Aquat. Sci.*, 37, 2300, 1980.

48. **Bender, M. E., Matson, W. R., and Jordan, R. A.**, On the significance of complexing agents in secondary sewage effluent, *Environ. Sci. Technol.*, 4, 520, 1970.

49. **Langis, R., Couture, P., de la Noüe, J., and Methot, N.**, Induced responses on algal growth and phosphate removal by three molecular weight DOM fractions from a secondary effluent, *J. Water Pollut. Control Fed.*, 58, 1073, 1986.

50. **De Kock, P. C.**, Influence of humic acids on plant growth, *Science*, 121, 473, 1955.

51. **Provasoli, L.**, Organic regulation of phytoplankton fertility, in *The Sea*, Hill, M. N., Ed., Interscience, London, 1963, 115.

52. **Sakamoto, M.**, Chemical factors involved in the control of phytoplankton production in the experimental lake area, Northwest Ontario, *J. Fish. Res. Board Can.*, 28, 203, 1971.

53. **Mahler, H. F. and Cordes, E. H.**, *Biological Chemistry*, Harper & Row, New York, 1966, 872.

54. **Lehninger, A. L.**, *Biochemistry*, 2nd ed., Worth Publishers, New York, 1975, 1104.

55. **Huntsman, S. A. and Sunda, W. G.**, The role of trace metals in regulating phytoplankton growth with emphasis on Fe, Mn and Cu, in *The Physiological Ecology of Phytoplankton*, Morris, I., Ed., University of California Press, Berkeley, 1980, 285.

56. **Bruland, K. W.**, Trace elements in seawater, in *Chemical Oceanography*, Vol. 8, Riley, D. and Chester, R., Eds., Academic Press, London, 1983, 157.

57. **Boyle, E. and Edmond, J. M.**, Copper in the surface waters south of New Zealand, *Nature*, 263, 107, 1975.

58. **Bruland, K. W., Franks, R. P., Knauer, G. A., and Martin, J. H.**, Sampling and analytical methods for the determination of copper, cadmium, zinc and nickel at the nanogram per liter level in seawater, *Anal. Chim. Acta*, 105, 233, 1979.

59. **Knauer, G. A. and Martin, J. H.**, Primary production and carbon-nitrogen fluxes in the upper 1500 m of the northeast Pacific, *Limnol. Oceanogr.*, 26, 181, 1981.

60. **Fitzwater, S. E., Knauer, G. A., and Martin, J. H.**, Metal contamination and primary production: field and laboratory methods of control, *Limnol. Oceanogr.*, 27, 544, 1982.

61. **Martin, J. H. and Knauer, G. A.**, VERTEX: manganese transport with $CaCO_3$, *Deep-Sea Res.*, 30, 411, 1983.

62. **Sunda, W. G. and Guillard, R. R. L.**, The relationship between cupric ion activity and the toxicity of copper to phytoplankton, *J. Mar. Res.*, 34, 511, 1976.

63. **Anderson, D. M. and Morel, F. M. M.**, Copper sensitivity of *Gonyaulax tamarensis*, *Limnol. Oceanogr.*, 23, 283, 1978.

64. **Anderson, D. M. and Morel, F. M. M.**, The influence on aqueous iron chemistry on the uptake of iron by the coastal diatom *Thalassiosira weissflogii*, *Limnol. Oceanogr.*, 27, 789, 1982.

65. **Sunda, W. G. and Hanson, A. K.**, Measurement of free cupric ion concentration in seawater by a ligand competition technique involving copper sorption onto C_{18} SEP-PAK cartridges, *Limnol. Oceanogr.*, 32, 537, 1987.

66. **van den Berg, C. M. G.**, Determination of copper complexation with natural organic ligands in seawater by equilibration with MnO_2. II. Experimental procedures and application to surface seawater, *Mar. Chem.*, 11, 323, 1982.

67. **van den Berg, C. M. G.**, Complexation of copper by organic material in the Irish Sea, *Mar. Chem.*, 14, 201, 1984.

68. **van den Berg, C. M. G.**, Determination of the complexation capacity and conditional stability constants of complexes of copper (II) with natural organic ligands in seawater by cathodic stripping voltammetry of copper-catechol complex ions, *Mar. Chem.*, 15, 1, 1984.

69. **van den Berg, C. M. G.**, Determination of the zinc complexation capacity in seawater by cathodic stripping voltammetry of zinc-APDC complex ions, *Mar. Chem.*, 16, 121, 1985.
70. **Waite, T. D. and Morel, F. M. M.**, Characterization of complexing agents in natural waters by copper (II)/copper (I) amperometry, *Anal. Chem.*, 55, 1268, 1983.
71. **Jackson, G. A. and Morgan, J. J.**, Trace metal-chelator interactions and phytoplankton growth in seawater media: theoretical analysis and comparison with reported observations, *Limnol. Oceanogr.*, 23, 268, 1978.
72. **Sunda, W. G., Engel, D. W., and Thuotte, R. M.**, Effect of chemical speciation on toxicity of cadmium to grass shrimp, *Palamonetes pugio:* importance of free cadmium ion, *Environ. Sci. Technol.*, 12, 409, 1978.
73. **Sunda, W. G. and Ferguson, R. L.**, Sensitivity of natural bacterial communities to additions of copper and to copper ion activity: a bioassay of copper complexation in seawater, in *Trace Metals in Sea Water*, Wong, C. S. and Boyle, E., Eds., Plenum Press, New York, 1983, 871.
74. **Tuschall, J. R. and Brezonik, P. L.**, Complexation of heavy metals by aquatic humus: a comparative study of five analytical methods, in *Aquatic and Terrestrial Humic Materials*, Christman, R. F. and Gjessing, E. T., Eds., Ann Arbor Science, Ann Arbor, MI, 1983, 275.
75. **Westall, J. C., Morel, F. M. M., and Hume, D. N.**, Chloride interference in cupric selective electrode measurements, *Anal. Chem.*, 51, 1792, 1979.
76. **Kononova, M. M.**, *Soil Organic Matter*, 2nd ed., Pergamon Press, New York, 1966.
77. **Gjessing, E. T.**, Humic substances in natural water: method for separation and characterization, in *Chemical Environment in the Aquatic Habitat*, Golterman, N. L. and Clymo, R. S., Eds., Elsevier/North-Holland, Amsterdam, 1967, 191.
78. **Kemp, A. L. W. and Mudrochova, A.**, The distribution and nature of amino acids and other nitro-containing compounds in Lake Ontario surface sediments, *Geochim. Cosmochim. Acta*, 37, 2191, 1973.
79. **Kemp, A. L. W. and Wong, H. K. T.**, Molecular-weight distribution of humic substances from lakes Ontario and Erie sediments, *Chem. Geol.*, 14, 15, 1974.
80. **Rashid, M. A. and King, L. A.**, Molecular weight distribution measurements on humic and fulvic fractions from marine clays on the Scotian Shelf, *Geochim. Cosmochim. Acta*, 33, 147, 1969.
81. **Rashid, M. A. and Prakash, A.**, Chemical characteristics of humic compounds isolated from some decomposed algae, *J. Fish. Res. Board Can.*, 29, 55, 1972.
82. **Neilands, J. B.**, Hydroxamic acids in nature, *Science*, 156, 1443, 1967.
83. **Murphy, T. P., Lean, D. R. S., and Nalewajko, C.**, Blue-green algae: their excretion of ion selective chelators enables them to dominate other algae, *Science*, 192, 900, 1976.
84. **McKnight, D. M. and Morel, F. M. M.**, Release of weak and strong copper-complexing agents by algae, *Limnol. Oceanogr.*, 24, 823, 1978.
85. **McKnight, D. M. and Morel, F. M. M.**, Copper complexation by siderochromes from filamentous blue-green algae, *Limnol. Oceanogr.*, 25, 62, 1980.
86. **Simpson, F. B. and Neilands, J. B.**, Siderochromes in Cyanophyceae: isolation and characterization of schizokinen from *Anabaena* sp., *J. Phycol.*, 12, 44, 1976.
87. **Jones, G. J., Palenik, B. P., and Morel, F. M. M.**, Trace metal reduction by phytoplankton: the role of plasmalemma redox enzymes, *J. Phycol.*, 23, 237, 1987.
88. **Harrison, G. I. and Morel, F. M. M.**, Antagonism between cadmium and iron in the marine diatom, *Thalassiosira weissflogii*, *J. Phycol.*, 19, 495, 1983.
89. **Sunda, W. G. and Huntsman, S. A.**, Effect of competitive interactions between manganese and copper on cellular manganese and growth in estuarine and oceanic species of the diatom *Thalassiosira*, *Limnol. Oceanogr.*, 29, 924, 1983.
90. **Morel, F. M. M. and Morel-Laurens, N. M. L.**, Trace metals and plankton in the ocean: facts and speculations, in *Trace Metals in Sea Water*, Wong, C. S. and Boyle, E., Eds., Plenum Press, New York, 1983, 841.
91. **Hewitt, E. J.**, Relation of manganese and some other metals to iron status in plants, *Nature*, 161, 489, 1949.
92. **Healy, W. B., Cheng, S.-C., and McElroy, W. D.**, Metal toxicity and iron deficiency effects on enzymes in *Neurospora*, *Arch. Biochem. Biophys.*, 54, 206, 1955.
93. **Enari, T. M. and Kauppinen, V.**, Interaction of cobalt and iron in the riboflavin production of *Candida guilliermondii*, *Acta Chem. Scand.*, 15, 1513, 1961.
94. **Muthukrishnan, S., Padmanaban, G., and Parma, P. S.**, Regulation of heme biosynthesis in *Neurospora crassa*, *J. Biol. Chem.*, 244, 4241, 1969.
95. **Shankar, K. and Bard, R. C.**, Effect of metallic ions on the growth, morphology and metabolism of *Clostridium perfringens*, *J. Bacteriol.*, 69, 444, 1955.
96. **Hamilton, D. L. and Valberg, L. S.**, Relationship between cadmium and iron absorption, *Am. J. Physiol.*, 227, 1033, 1974.
97. **Davies, A. G.**, Iron, chelation and the growth of marine phytoplankton. I. Growth kinetics and chlorophyll production in cultures of the euryhaline flagellate *Dunaliella tertiolecta* under iron-limiting conditions, *J. Mar. Biol. Assoc. U.K.*, 50, 65, 1970.

98. **Cardenas, J., Rivas, J., Paneque, A., and Losada, M.,** Effect of iron supply on the activities of the nitrate-reducing system from *Chlorella,* in *Bioenergetics and Metabolism of Green Algae,* Vol. 1, Cardenas, J., Ed., MSS Information, New York, 1974, 10.
99. **Glover, H.,** Effects of iron deficiency on *Isochrysis galbana* (Chlorophyceae) and *Phaeodactylum tricornutum* (Bacillariophyceae), *J. Phycol.,* 13, 208, 1977.
100. **Goering, J. J., Boisseau, D., and Hattori, A.,** Effects of copper on silicic acid by a marine phytoplankton population: controlled ecosystem pollution experiment, *Bull. Mar. Sci.,* 27, 58, 1977.
101. **Harrison, W. G., Eppley, R. W., and Renger, E. H.,** Phytoplankton nitrogen metabolism, nitrogen budgets, and observations on copper toxicity: controlled ecosystem pollution experiment, *Bull. Mar. Sci.,* 27, 44, 1977.
102. **Morel, N. M. L. and Morel, F. M. M.,** Lag Phase Promotion in the Growth in *Pyramimonas* 1 by Manipulation of the Trace Metal Chemistry of the Medium, Tech. Note No. 17, R. M. Parsons Laboratory, Massachusetts Institute of Technology, Cambridge, 1976, 29.
103. **Mueller, B.,** Some Aspects of Iron Limitation in a Marine Diatom, Master's thesis, University of British Columbia, Vancouver, 1985.
104. **Rueter, J. G., Jr. and Morel, F. M. M.,** The interaction between zinc deficiency and copper toxicity as it affects the silicic acid uptake mechanisms in *Thalassiosira pseudonana, Limnol. Oceanogr.,* 26, 67, 1981.
105. **Rueter, J. G., Jr., Chisholm, S. W., and Morel, F. M. M.,** Effects of copper toxicity on silicic acid uptake and growth in *Thalassiosira pseudonana, J. Phycol.,* 17, 270, 1981.
106. **Rueter, J. G. and Ades, D. R.,** The role of iron nutrition in photosynthesis and nitrogen assimilation in *Scenedesmus quadricauda* (Chlorophyceae), *J. Phycol.,* 23, 452, 1987.
107. **Verstreate, D. R., Storch, T. A., and Dunham, V. L.,** A comparison of the influence of iron and nitrate metabolism of *Anabaena* and *Scenedesmus, Physiol. Plant.,* 50, 47, 1980.
108. **Kamp-Nielsen, L.,** The influence of copper on the photosynthesis and growth of *Chlorella pyrenoidosa, Dan. Tidsskr. Farm.,* 43, 249, 1969.
109. **Paskins-Hurlburt, A., Tanaka, Y., and Skoryna, S. C.,** Isolation and metal binding properties of fucoidan, *Bot. Mar.,* 19, 327, 1976.
110. **Veroy, R. L., Montano, N., de Guzman, M. L. B., Laserna, E. C., and Cajipe, G. J. B.,** Studies on the binding of heavy metals to algal polysaccharides from Philippine seaweeds. I. Carrageenan and binding of lead and cadmium, *Bot. Mar.,* 23, 59, 1980.
111. **Skoryna, S. C. and Tanaka, Y.,** Biological activity of fractionation products of brown marine algae, in *Proc. 6th Int. Seaweed Symp.,* Margalef, R., Ed., Direccion General de Pesca Maritima, Madrid, 1969, 737.
112. **Bryan, G. W. and Hummerstone, L. G.,** Brown seaweed as an indicator of heavy metals in estuaries in south-west England, *J. Mar. Biol. Assoc. U.K.,* 53, 705, 1973.
113. **Woolston, M. E., Breck, W. G., and VanLoon, G. W.,** A sampling study of the brown seaweed, *Ascophyllum nodosum* as a marine monitor for trace metals, *Water Res.,* 16, 687, 1982.
114. **Greene, B., Hosea, M., McPherson, R., Henzl, M., Alexander, M. D., and Darnall, D. W.,** Interaction of gold(I) and gold(III) complexes with algal biomass, *Environ. Sci. Technol.,* 20, 627, 1986.
115. **Darnall, D. W., Greene, B., Henzl, M. T., Hosea, J. M., McPherson, R., and Alexander, M. D.,** Selective recovery of gold and other metal ions from an algal biomass, *Environ. Sci. Technol.,* 20, 206, 1986.
116. **Hosea, M., Greene, B., McPherson, R., Henzl, M., Alexander, D., and Darnall, D. W.,** Interaction of gold(I) and gold(III) complexes with algal biomass, *Inorg. Chim. Acta,* 123, 161, 1986.
117. **Greene, B., Henzl, M. T., Hosea, J. M., and Darnall, D. W.,** Elimination of bicarbonate interference in the binding of U(VI) in mill-waters to freeze-dried *Chlorella vulgaris, Biotechnol. Bioeng.,* 28, 764, 1986.
118. **Kagi, J. H. R. and Nordberg, N., Eds.,** *Metallothionein,* Birkhauser, Basel, 1979.
119. **Hamer, D. H.,** Metalliothionein, *Annu. Rev. Biochem.,* 55, 913, 1986.
120. **Foulkes, E. C., Ed.,** *Biological Roles of Metallothionein,* Elsevier, New York, 1982.
121. **Karim, M.,** Metallothioneins: proteins in search of function, *Cell,* 41, 9, 1985.
122. **Butt, T. R. and Ecker, D. J.,** Yeast metallothionein and applications in biotechnology, *Microbiol. Rev.,* 51, 351, 1987.
123. **Grill, E., Winnacker, E. L., and Zenk, M. H.,** Phytochelatins, a class of heavy-metal-binding peptides from plants, are functionally analogous to metallothioneins, *Proc. Natl. Acad. Sci. U.S.A.,* 84, 439, 1987.
124. **Grill, E., Winnacker, E. L., and Zenk, M. H.,** Synthesis of seven different homologous phytochelatins in metal-exposed *Schizosaccharomyces pombe* cells, *FEBS Lett.,* 197, 115, 1986.
125. **Dayton, L. and Lewin, R. A.,** The effects of lead on algae. III. Effects of Pb on population growth curves in two-membered cultures of phytoplankton, *Arch. Hydrobiol. Suppl.,* 49, 25, 1975.
126. **Sunda, W. G.,** Relationship between Cupric Ion Activity and Toxicity of Copper to Phytoplankton, Ph.D. thesis, Massachusetts Institute of Technology, Cambridge, 1975.

127. **Stoltzberg, R. J. and Rosin, D.,** Chromatographic measurement of submicromolar strong complexing capacity in phytoplankton media, *Anal. Chem.,* 49, 226, 1977.
128. **Gnassia-Berelli, M., Romeo, M., Laumond, F., and Pesando, D.,** Experimental studies on the relationship between natural copper complexes and their toxicity to phytoplankton, *Mar. Biol.,* 47, 15, 1978.
129. **Biggs, D. C., Powers, C. D., Rowland, R. G., O'Connors, H. B., and Wurster, C. F.,** Uptake of polychlorinated biphenyls by natural phytoplankton assemblages: field and laboratory determination of ^{14}C-PCB particle-water index of sorption, *Environ. Pollut. Ser. A,* 22, 101, 1980.
130. **Davis, E. M., Turley, J. E., Casserly, D. M., and Guthrie, R. K.,** Partitioning of selected organic pollutants in aquatic ecosystems, in *Biodeterioration 5: Papers Presented at the 5th International Biodeterioration Symposium,* Oxley, T. A. and Barry, S. M., Eds., John Wiley & Sons, New York, 1982, 176.
131. **Sodergren, A.,** Transport, distribution, and degradation of chlorinated hydrocarbon residues in aquatic model ecosystems, *Oikos,* 24, 30, 1973.
132. **Butler, G. L.,** Algae and pesticides, *Residue Rev.,* 66, 19, 1977.
133. **O'Kelley, J. C. and Deason, T. R.,** Degradation of Pesticides by Algae, EPA-600/3-76-022, National Technical Information Services, Springfield, VA, 1976.
134. **Pahdy, R. N.,** Cyanobacteria and pesticides, *Residue Rev.,* 95, 1, 1985.
135. **Wright, S. J. L.,** Interactions of pesticides with microalgae, in *Pesticide Microbiology,* Hill, I. R. and Wright, S. J. L., Eds., Academic Press, London, 1978, 585.
136. **Neudorf, S. and Khan, M. A. Q.,** Pick-up and metabolism of DDT, dieldrin and photodieldrin by a fresh water alga *(Ankistrodesmus amalloides)* and a microcrustacean *(Daphnia pulex), Bull. Environ. Contam. Toxicol.,* 13, 443, 1975.
137. **Kruglov, Y. V. and Paromenskaja, L. N.,** Detoxication of simazine by microscopic algae, *Mikrobiologiya,* 39, 157, 1970.
138. **Maguire, R. J., Wong, P. T. S., and Rhamey, J. S.,** Accumulation and metabolism of tri-n-butyltin cation by a green alga, *Ankistrodesmus falcatus, Can. J. Fish. Aquat. Sci.,* 41, 537, 1984.
139. **Ellis, B. E.,** Degradation of phenolic compounds by freshwater algae, *Plant Sci. Lett.,* 8, 213, 1977.
140. **Cerniglia, C. A., Gibson, D. T., and Van Baalen, C.,** Oxidation of naphthalene by cyanobacteria and microalgae, *J. Gen. Microbiol.,* 116, 495, 1980.
141. **Craigie, J. S., McLachlan, J., and Towers, G. H. N.,** A note on the fission of an aromatic ring by algae, *Can. J. Bot.,* 43, 1589, 1965.
142. **Bowes, G. W.,** Uptake and metabolism of 2,2-bis(p-chlorophenyl)-1,1,1-trichloroethane (DDT) by marine phytoplankton and its effect on growth and chloroplast electron transport, *Plant Physiol.,* 49, 172, 1972.
143. **Patil, K. C., Matsumura, F., and Boush, G. M.,** Metabolic transformation of DDT, dieldrin, aldrin, and endrin by marine microorganisms, *Environ. Sci. Technol.,* 6, 629, 1972.
144. **Valentine, J. P. and Bingham, S. W.,** Influence of several algae on 2,4-D residues in water, *Weed Sci.,* 22, 358, 1974.
145. **Rice, C. P. and Sikka, H. C.,** Uptake and metabolism of DDT by six species of marine algae, *Agric. Food Chem.,* 21, 148, 1973.
146. **Keil, J. E. and Priester, L. E.,** DDT uptake and metabolism by a marine diatom, *Bull. Environ. Contam. Toxicol.,* 4, 169, 1969.
147. **Vose, J. R., Cheng, J. Y., Antia, N. J., and Towers, G. H. N.,** The catabolic fission of the aromatic ring of phenylalanine by marine planktonic algae, *Can. J. Bot.,* 49, 259, 1971.
148. **Cerniglia, C. E., Gibson, D. T., and Van Baalen, C.,** Naphthalene metabolism by diatoms isolated from the Kachemak Bay region of Alaska, *J. Gen. Microbiol.,* 128, 987, 1982.
149. **Miyazaki, S. and Thorsteinson, A. J.,** Metabolism of DDT by fresh water diatoms, *Bull. Environ. Contam. Toxicol.,* 8, 81, 1972.
150. **Cerniglia, C. E., Gibson, D. T., and Van Baalen, C.,** Algal oxidation of aromatic hydrocarbons: formation of 1-naphthol from naphthalene by *Agmenellum quadruplicatum,* strain PR-6, *Biochem. Biophys. Res. Commun.,* 88, 50, 1979.
151. **Cerniglia, C. E., Freeman, J. P., and Van Baalen, C.,** Biotransformation and toxicity of aniline derivatives in cyanobacteria, *Arch. Microbiol.,* 130, 272, 1981.
152. **Cerniglia, C. E., Freeman, J. P., Althaus, J. R., and Van Baalen, C.,** Metabolism and toxicity of 1- and 2-methylnaphthalene and their derivatives in cyanobacteria, *Arch. Microbiol.,* 136, 177, 1983.
153. **Matsumura, F. and Esaac, E. G.,** Degradation of pesticides by algae and aquatic organisms, in *Pesticide and Xenobiotic Metabolism in Aquatic Organisms,* ACS Symp. Ser. No. 99, Khan, M. A. Q., Lech, J. J., and Menn, J. J., Eds., American Chemical Society, Washington, D.C., 1979, 371.
154. **Cerniglia, C. E., Van Baalen, C., and Gibson, D. T.,** Metabolism of naphthalene by the cyanobacterium *Oscillatoria* sp., strain JCM, *J. Gen. Microbiol.,* 116, 485, 1980.
155. **Cerniglia, C. E., Van Baalen, C., and Gibson, D. T.,** Oxidation of biphenyl by the cyanobacterium, *Oscillatoria* sp., strain JCM, *Arch. Microbiol.,* 125, 203, 1980.
156. **Khalil, Z. and Mostafa, I. Y.,** Interactions of pesticides with freshwater algae. I. Effect of methomyl and its possible degradation by *Phormidium fragile, J. Environ. Sci. Health B,* 21, 289, 1986.

157. **Hill, I. R.,** Microbial transformation of pesticides, in *Pesticide Microbiology,* Hill, I. R. and Wright, S. J. L., Eds., Academic Press, London, 1978, 187.

158. **Brink, R. H.,** Biodegradation of organic chemicals in the environment, in *Environmental Health Chemistry,* McKinney, J. D., Ed., Ann Arbor Science, Ann Arbor, MI, 1981, 75.

159. **Narro, M. L., Cerniglia, C. E., Gibson, D. T., and Van Baalen, C.,** The oxidation of aromatic compounds by microalgae, in *Environmental Regulation of Microbial Metabolism,* Kulaev, I. S., Dawes, E. A., and Tempest, D. W., Eds., Academic Press, London, 1985.

160. **Hollister, T. A., Walsh, G. E., and Forester, J.,** Mirex and marine unicellular algae: accumulation, population growth and oxygen evolution, *Bull. Environ. Contam. Toxicol.,* 14, 753, 1975.

161. **Sodergren, A.,** Accumulation and distribution of chlorinated hydrocarbons in cultures of *Chlorella pyrenoidosa* (Chlorophyceae), *Oikos,* 22, 215, 1971.

162. **Sodergren, A.,** Uptake and accumulation of ^{14}C-DDT by *Chlorella* sp. (Chlorophyceae), *Oikos,* 19, 126, 1968.

163. **Veber, K., Zahradnik, J., Breyl, I., and Kredl, R.,** Toxic effect and accumulation of atrazine in algae, *Bull. Environ. Contam. Toxicol.,* 27, 872, 1981.

164. **Butler, G. L., Deason, T. R., and O'Kelley, J. C.,** Loss of five pesticides from cultures of twenty-one planktonic algae, *Bull. Environ. Contam. Toxicol.,* 13, 149, 1975.

165. **Gregory, W. W., Reed, J. K., and Forester, L. E.,** Accumulation of parathion and DDT by some algae and protozoa, *J. Protozool.,* 16, 69, 1969.

166. **Glooschenko, V., Holdrinet, M., Lott, J. N. A., and Frank, R.,** Bioconcentration of chlordane by the green alga *Scenedesmus quadricauda, Bull. Environ. Contam. Toxicol.,* 21, 515, 1979.

167. **Keil, J. E., Priester, L. E., and Sandifer, S. H.,** Polychlorinated biphenyl (Aroclor 1242): effects of uptake on growth, nucleic acids, and chlorophyll of a marine diatom, *Bull. Environ. Contam. Toxicol.,* 6, 156, 1971.

168. **Cox, J. L.,** Low ambient level uptake of ^{14}C-DDT by three species of marine phytoplankton, *Bull. Environ. Contam. Toxicol.,* 5, 218, 1970.

169. **Mukerji, F., Glover, H. E., and Morris, I.,** Diversity in the mechanism of carbon dioxide fixation in *Dunaliella tertiolecta* (Chlorophyceae), *J. Phycol.,* 14, 137, 1978.

170. **Glover, H. E. and Morris, I.,** Photosynthesis carboxylating enzymes in marine phytoplankton, *Limnol. Oceanogr.,* 24, 510, 1979.

171. **Li, W. K. W., Glover, H. E., and Morris, I.,** Physiology of carbon photoassimilation by *Oscillatoria thiebautii* in the Caribbean Sea, *Limnol. Oceanogr.,* 25, 447, 1980.

Chapter 8

ALGAL CULTURE SYSTEMS

Mark E. Huntley, Arthur M. Nonomura, and Joël de la Noüe

TABLE OF CONTENTS

I. INTRODUCTION

In agriculture, algae are ideal for treatment of drainage, primarily because they are plants with the same fundamental requirements as the cash crops from which wastewaters originate. Indeed, agricultural drainage contains many undesirable substances, such as nitrates, which often support blooms of naturally occurring algae.[1] Algae have long been employed in process streams for civic liquid waste treatment,[2-7] in sewage ponds associated with livestock feeds,[8] and as nutrient scrubbers in specialized mariculture systems.[9,10] However, controlled cultures of algae have not been widely used for the treatment of agricultural drainage. In this chapter, we present an overview of some culture system technologies, ranging in stage of development from well tested to experimental, which may be suitable for the treatment of agricultural wastewater. Furthermore, algal harvesting methodologies are reviewed, some of the economic by-products of algal mass culture are discussed, and information is provided on where to obtain "starter" cultures.

Although there exist approximately 20,000 species of algae, including the cyanobacteria, barely 100 of these have been sufficiently studied by phycologists to qualify as "known". It is generally recognized that algae exist in many forms, from unicellular microalgae only a few micrometers in diameter to frondose kelps many meters in height.[11,12] Many of these have been mass cultured in systems as diverse as open ponds,[13] plastic bags and tubes,[14,15] turfs,[16] and maricultured rope networks.[17] Being nature's principal conduit for the transformation of solar energy and dissolved substances into living matter, algae come equipped with a broad range of physiological specializations and metabolic pathways[18-20] that enable them to thrive in water high in dissolved nitrogen, phosphorus, organic matter, trace metals, and even xenobiotic compounds — all of which may occur in agricultural drainage.[21]

The overall process of mass culture of algae involves the introduction of nutrient medium into a growth environment, production of the algae, and subsequent harvesting or removal of the algae from the spent medium. All mass culture programs must devise methods to efficiently accomplish each of these processes. In evaluating whether algae should or should not be applied to certain drainage problems, the limitations of these organisms should be considered.[22] Some species, especially the large marine forms, grow slowly, and although many microalgal species double their biomass within a day, their growth rates are not as rapid as those of most bacteria. The complex life histories of some algae may necessitate sophisticated culture methods. As cultures age and reach high population densities, auto-regulatory mechanisms may stop the growth of some species. The possible need for nutrient supplementation may require not only the cost of additional nutrients and chelators, but also the added cost of periodic monitoring. Like bacteria, microalgae in suspension will contribute to the total suspended solids unless they are harvested or otherwise removed. In large culture systems, settling of algal cells must be avoided to prevent decomposition and subsequent anoxia; therefore, energy-requiring physical circulation mechanisms are necessary. Finally, climatic factors must be regulated (either by choice of location or by its control) to optimize conditions of temperature, radiant solar energy, wind, predators, and influent water quality.

II. OPEN CULTURE

Open culture systems are characterized as being open to the atmosphere. They may take the form of ponds, channels, or raceways in which environmental control is limited. Due largely to their relative simplicity, in an engineering sense, open systems were the first mass culture devices to receive considerable attention.

A. MULTISPECIES OPEN CULTURE

Early attempts at the mass culture of microalgae utilizing *Chlorella* were able to produce yields of up to 11 g dry weight/m^2/d on a short-term basis in a small-volume (~750 l) growth apparatus,[23] but long-term production yields were a factor of five lower. During the 1950s, research was conducted on the production of algae in a large pond utilizing domestic wastewater as the nutrient source.[7,13,24] This system, located in Richmond, CA, yielded 12 to 18 g dry weight/m^2/d[25,26] and has served as the essential model for many subsequent mass culture efforts.

Most mass culture operations over the past 40 years have involved shallow, open ponds mixed by a variety of means, including pumps, paddle wheels, or aeration. The depth of the pond is usually 10 to 40 cm, and its surface area generally ranges from 100 to 10,000 m^2. Open ponds are relatively inexpensive with regard to materials. In the simplest form, the pond consists of a uniformly deep bottom surface surrounded by levees and separated into two channels by a levee which partially bisects the pond. For large ponds, it is advisable to grade the bottom with the assistance of laser leveling technology. The pond may or may not be lined, but in any case a smooth surface is desirable; this both reduces energy losses due to friction and minimizes the amount of surface area onto which algae may settle and decompose. Ponds have been lined with plastic, finely crushed gravel, clay, concrete, and even with the cultured algae themselves, which may produce extracellular polysaccharides that create a slimy, smooth surface.[27]

Algae in most open pond systems are usually mixtures of endemic species. Inoculating a large volume of water with a single species requires special maintenance procedures and conditions (see Section II.B below). The method which is the least expensive and has the lowest maintenance requirements consists of enhancing the growth of algae which occur naturally in local waters. With changes in season and water quality, species composition may shift or change completely. The genera in Table 1 are those typically observed in sanitary treatment waters, and they are classified according to the conditions in which they are most likely to be found. Given the extremely limited experience with algal mass culture in agricultural wastewater, it is difficult to predict whether these same species will prevail. The points we wish to emphasize are that (1) a relatively small group of species will tend to dominate open cultures at any time at a given location and that (2) species composition may change dramatically, and often unpredictably, on very short time scales (on the order of days).

Economic treatment of agricultural wastes has been undertaken with "advanced integrated ponds".[27] These systems were originally proposed as an alternative to mechanical primary and secondary treatment processes in the climatically stable sun belt of North America. Where long-term freezing does not occur, such ponds may be ideal, especially in the temperate to tropical climes of developing countries. Low- to medium-cost systems can be designed up to tertiary treatment.

Primary treatment involves the separation of solids. Advanced integrated ponds are designed with sedimentation and fermentation pits to enhance the activities of methane bacteria. Algae in the surface waters of buffer ponds raise the pH, temperature, and oxygen content. The high pH contributes to the precipitation of heavy metals and the reduction of coliform counts.

Secondary treatment is achieved with a high-rate pond which exploits the oxidative capabilities of algae. Algae produce oxygen photosynthetically and, in the process, raise waters above pH 9. The combination of high pH and high dissolved oxygen concentration enhances ammonia outgassing and induces precipitation of calcium phosphate; it also enhances disinfection and metal precipitation. The growth rate of the algae can be controlled by adjusting pond depth and residence time. This control can be used to maintain optimum uptake of dissolved nutrients such as nitrate and phosphate.

Tertiary treatment involves the removal of the cultured algae. Methods of separation are discussed in detail in this chapter (see Section IV). The water can be treated to within acceptable limits of suspended solids and very low biological oxygen demand (BOD), and

TABLE 1

Microalgal Genera Typically Observed in Urban Wastewater Systems, Characterized According to Bad Smell and/or Taste and Filter Clogging or Sealing Ability

Species	Characteristic		
	Bad taste or smell	Filter clogging	Reservoir sealant
Cyanophyta			
Anabaena	•	•	
Anacystis	•		•
Aphanizomenon	•		
Chroococcus		•	
Coelosphaerium	•		
Gloeocapsa	•		
Gomphosphaeria			•
Lyngbya			•
Microcystis	•		
Oscillatoria	•	•	
Phormidium			•
Rivularia	•	•	
Schizothrix			•
Spirulina			•
Tolypothrix			•
Chlorophyta			
Chlorella		•	
Cladophora			•
Closterium		•	
Dunaliella		•	•
Gloeocystis	•		
Hydrodictyon	•		
Pandorina	•		
Rhizoclonium			•
Scenedesmus	•		
Staurastrum	•		
Spirogyra		•	
Volvox			b
Chrysophyta			
Dinobryon	•		
Mallmonas	•		
Synura	•		
Uroglenopsis	•		
Cryptophyta			
Cryptomonas			•
Charophyta			
Nitella	•		
Xanthophyta			
Tribonema		•	
Bacillariophyta			
Amphora			•
Asterionella	•	•	
Cyclotella	•	•	
Cymbella		•	
Fragilaria	•	•	
Melosira		•	
Navicula		•	•
Nitzschia			•
Synedra	•	•	•
Tabellaria	•	•	

TABLE 1 (continued)
Microalgal Genera Typically Observed in Urban Wastewater
Systems, Characterized According to Bad Smell and/or Taste
and Filter Clogging or Sealing Ability

Species	Characteristic		
	Bad taste or smell	Filter clogging	Reservoir sealant
Dinophyta			
Amphidinium			•
Ceratium	•		
Gymnodinium			•
Peridinum	•		•

Note: Data compiled from Davis,[137] Tchobanoglous and Schroeder,[155] and experience of the authors.

it may then be reused. The harvested algae may be used as feedstock for methane production or processed for other commercial uses. Treatment of agricultural drainage probably would utilize only secondary and tertiary treatments, with high-rate ponds initiated directly from the drain stream.

B. UNIALGAL OPEN CULTURE

The culture of a single species in an open system is accomplished by rendering conditions in the system so hostile that only select species can thrive. The principal means of achieving control are (1) the use of a species with a very high growth rate and (2) unusual temperature, pH, or nutrient conditions. Early attempts at mass culture focused primarily on *Scenedesmus* and *Chlorella* spp. These species could be maintained in predominantly unialgal cultures primarily because of their high growth rates, which allow them to compete favorably with any potential contaminant species. Nevertheless, their optimal growth environments are sufficiently similar to those of many other species that, should the culture conditions fluctuate enough to reduce their growth rates, the culture can be rapidly invaded by competitive species.

The two principal genera which have been favored for commercial production are *Dunaliella* and *Spirulina*. These can be produced economically from virtually unialgal cultures because they can withstand conditions which are not tolerated by most microalgae. *Dunaliella* spp. grow well in highly saline waters and produce large quantities of glycerol, a valuable by-product, in salt solutions which are three to four times greater than natural seawater.[28-30] Under these conditions, competition from other organisms is eliminated. *Spirulina* spp. can be maintained successfully in unialgal culture because they grow optimally at high temperatures (~30°C),[31,32] tolerate elevated salinity,[32,33] and thrive in waters with a pH range of 9 to 10,[32-34] whereas many other microalgae prefer a pH closer to 7.

For application to treating agricultural drainage, both *Dunaliella* and *Spirulina* spp. might be worthy of consideration, particularly in saline and alkaline waters. However, in the absence of the special conditions preferred by these species, other approaches should be considered.

C. RACEWAYS

Raceways, as their name implies, involve systems in which the culture medium is mixed or propelled at rates which are sufficient to create a turbulent flow regime. Laboratory-scale cultures in Couette devices[36-38] and "algatrons"[39,40] clearly have demonstrated that photosynthetic efficiencies can be markedly increased through turbulent mixing, which presumably provides the environment necessary to take advantage of the "flashing light effect", which is commonly known to enhance photosynthesis.[41-43]

Perhaps the best example of the raceway design is provided by the recent work of Laws and co-workers.[44,45] In a 48-m^2 system containing 4150 l circulating at a rate of 30 cm/s by means of an airlift system and with an operating depth of ~8 cm, they achieved sustained production rates for the marine diatom *Phaeodactylum tricornutum* of >23 g dry weight/m^2/d. The system was operated out-of-doors in Hawaii, where ambient temperatures often exceeded the optimum for *P. tricornutum* (20°C);[46] Laws et al.[44] suggested that, under optimal temperature conditions, their system could have achieved production rates of 36 g dry weight/m^2/d — approximately double that attainable in comparable open pond systems.

Raceways are more costly than open ponds, primarily due to the costs associated with producing a high flow rate. Not only are aeration systems more costly to construct and operate than the paddle wheels commonly used in open ponds, but additional costs are incurred by the materials required for providing low-friction surfaces within the raceway.

D. ADVANTAGES AND DISADVANTAGES

The principal advantages of open systems are the relatively low capital costs and the ease of operation. For regions where the water supply is consistent with respect to chemical composition and flow rate and where temperatures do not fluctuate greatly, open ponds or raceways may provide a suitable environment for treatment of agricultural drainage. Even under such relatively stable conditions, one should expect a multispecies culture. Maintenance of a unialgal culture may require extreme physical or chemical modifications to the environment and will impose strict limitations on the number of species which can be cultured.

The principal disadvantages of open systems arise from the difficulties involved in environmental control. Because the system is open to the atmosphere, the potential for contamination by unwanted species is always very high. The lack of control of other environmental variables, notably temperature and water chemistry, only serves to compound the potential for contamination. If there exists a need to cultivate specific species with one or more of the unique physiological or metabolic characteristics required for biodegradation (discussed in this book by Redalje et al.[18]), then open systems will probably not be sufficient to meet one's requirements.

III. CLOSED CULTURE

Closed culture systems are characterized as being closed to the atmosphere. The principal goal of closed culture is to increase the amount of environmental control. The mere act of separating the culture from the atmosphere provides the immediate benefit of eliminating the introduction of airborne contaminant species, but it also provides the basis for increasingly sophisticated control of temperature, light, nutrients, and gases. Closed mass-culture systems began to be suggested in various forms only after it became evident that open systems could not always be controlled by the introduction of fast-growing species. By virtue of their increased attention to environmental control, closed systems tend to be more costly than comparable open systems, and for many purposes costs are prohibitive. However, with continued improvement of the bioconcentrative and biodegradative abilities of selected algal strains, closed systems are likely to become more efficient (and hence affordable) for the treatment of agricultural and other wastewaters.

The lagging development of microalgal culture technology behind that of yeasts and bacteria stems from the fact that industry has exploited the latter organisms as transforming agents because of their high division rates and extracellular production of many desired products. Indeed, microalgae present a spectrum of problems, including (1) long generation times, (2) difficulty in harvesting, (3) requirement for light (which demands a different design for bioreactors or large pond areas for intensive culture), (4) lower active biomass concentration, and (5) less well-known physiology. Some of these difficulties are inherent in the biology of microalgae and will be difficult to overcome; however, others may be resolved by both technological and biotechnological advances.

Two categories of closed systems may be identified according to whether the algae are allowed to move freely in the culture suspension (mobilized algae) or if they are immobilized with respect to the motion of the culture medium (immobilized algae).

A. MOBILIZED ALGAE

One of the earliest versions of the closed mass-culture system was proposed by Dewey,[47] who suggested that the culture be pumped through a system of translucent tubes embedded in shallow trenches. This design eliminated the possibility of atmospheric contamination, but did not provide for the straightforward control of other environmental variables.

Many subsequent designs for closed mass-culture systems have employed essentially the same concept, i.e., a recirculating culture enclosed in translucent tubes,[48-51] but there also exist designs for enclosed raceways.[52] In each of these systems, the amount of additional environmental control is variable. The 5-m^3 system developed by Balloni et al.[49] had no provision for any additional controls; the culture was recirculated at 30 to 50 cm/s through flexible polyethylene tubing 14 cm in diameter and, over a period of several months, provided an average yield of 26 g/m^2/d of photosynthetic rhodobacteria and 11 g/m^2/d of *Spirulina maxima*.

The system developed by Gudin[48] did provide for temperature control. A network of adjacent plastic tubes was immersed in water. Heating was accomplished by solar energy. Cooling was achieved by regulating the depth to which the system was immersed via increasing or decreasing the amount of gas in the system, and a method was suggested for doing this automatically via a feedback system involving temperature sensors and controlled carbon dioxide addition. Field tests[15] of a 1-m^2 bioreactor realized a production of *Scenedesmus acutus* of 21 g/m^2/d over a period of several months, with a maximum of 34 g/m^2/d.

The tubular culture system designed by Huntley et al.[51] achieves temperature control by atmospheric regulation, which is accomplished by placing the entire culture system in a solar-heated plastic enclosure; evaporative cooling is employed for temperature reduction. All of the critical variables necessary for achieving optimum growth of the algae, including temperature, pH, nutrient concentrations, and biomass, as well as the harvesting cycle are microprocessor controlled. With the possibility for providing optimal growth conditions in this environment, it was estimated that one could attain productivity in the range of 40 to 65 g/m^2/d, more than double that attainable in conventional open pond systems.

The enclosed raceway system developed by Raymond[52] contains elaborate controls for pH, nutrient concentrations, and light transmission, as well as temperature. In tests of this apparatus with the marine species *Phaeodactylum tricornutum*, yields of 28 g/m^2/d were achieved. Raymond suggested that, with further optimization, this production could be doubled in certain locations and climatic conditions.

Closed systems present the possibility for using hyperconcentrated algal cultures and taking advantage of the heterotrophic/mixotrophic nutritional mode which can develop in such cultures. Until recently, most studies involving the mass culture of microalgae have been conducted with culture densities of around 150 to 200 mg dry weight/l, with the exception of a few studies where the algal concentration has been raised to ~1 g dry weight/l.[53,57] Lavoie and de la Noüe[58] recently investigated the behavior of so-called hyperconcentrated cultures, in which the biomass concentration of *Scenedesmus obliquus* was raised to 2.6 g dry weight/l. The use of hyperconcentrated suspensions has also been explored with immobilized microalgae, with equal success.[59] The rates of ammonium and phosphate removal from wastewaters are greatly increased in such cultures. The operation of these systems raises important technological problems for large-scale application at present, but they may prove applicable in the near future, especially for highly diluted effluents for which a high hydraulic dilution rate is required. Recirculation of the biomass is required to avoid a rapid washout of the culture; this could be accomplished with chemical flocculation, but the cost of flocculants is currently prohibitive. A great deal more research will be required before the use of hyperconcentrated cultures will become economical.

There is considerable evidence that the conditions of light limitation which prevail in hyperconcentrated cultures induce a heterotrophic or mixotrophic nutritional mode in the microalgae being cultured. Heterotrophy has been reported in microalgae, especially in diatoms,[60,61] but also in other groups.[62,67] The growth rate of heterotrophic algae is generally lower than that of photoautotrophs,[68,69] and the required carbon substrates are probably not present in most agricultural drainage. However, in special situations where drainage waters could be enriched with effluents of high organic load, heterotrophy might be profitably exploited, especially in conjunction with hyperconcentrated cultures. Research along these lines is, however, just beginning.

B. IMMOBILIZED ALGAE

This type of closed system depends upon using various techniques to fix algal cells or cellular organelles to substrates or matrices over or through which the growth medium is passed. In theory, immobilization offers many advantages.[70,71] Culture monospecificity can be maintained more easily, and the active biomass can be reused, which compensates for low algal growth rates (by comparison to bacteria) and allows for the operation of continuous systems. Higher cell concentrations can be maintained, therefore providing higher reaction rates. In contrast to mobilized algae systems, bioreactors can be operated at dilution rates that exceed the maximum specific growth rate, μ_{max}, without washout. An insoluble immobilization matrix may provide protection for aging cells (and thus enhance photosynthetic activity)[72] and, by analogy with higher plant cells, may lead to increased metabolite production.[73] Immobilization can provide increased temperature stability[74] (especially at low temperatures),[75] better and easier control of processes,[71] less demand for expensive bioreactors,[71] easier recovery of by-products,[71] and, finally, the possibility of incorporating efficient symbiotic bacteria.[76-78]

The attractive features of immobilization are still at the research level, however. One of the major problems is that the polymeric matrices used so far for immobilization, such as alginate,[79] agarose,[80] carrageenan,[59,78,81] and agar,[82], are too expensive to allow large-scale operations. Some specific problems arise with alginate: the matrix dissolves in phosphate-containing media, a situation likely to exist in agricultural drainage and, moreover, difficult to avoid with microalgae, since they require phosphate as a nutrient. Other polymers used for immobilization, such as polyacrylamide,[83] have proved to be toxic to living cells. One promising substance is chitosan, which has been used to successfully immobilize the cyanobacterium *Phormidium* sp.;[84] although chitosan is nontoxic, even the industrial grade is still rather expensive (~$6/lb, U.S. 1987).

New and cheaper immobilization materials will have to be tested before immobilized microalgae can be efficiently used for applied purposes. Until now relatively few systems using microalgae or cyanobacteria have been explored, but there is already considerable interest in many applications, including the production of polysaccharides,[85] fuels and chemicals,[86,87] hydrogen,[88] oxygen,[77,89] and the fixation of atmospheric nitrogen.[90] Removal of nutrients from domestic wastewater has also been studied.[81,84]

Using the small (4 × 12 μm) green microalga *Scenedesmus obliquus* grown on a secondary-treated urban effluent (N-NH$_4^+$, 8.0 to 16.8 mg/l; P-PO$_4^{3-}$, 2.3 to 3.4 mg/l; initial pH, 7.1 to 7.7), Chevalier and de la Noüe[81] showed that the removal of N-NH$_4^+$ and P-PO$_4^{3-}$ proceeded with similar kinetics for both free and immobilized cells cultured in batch mode. The same efficiency for ammonium and phosphate removal by free and immobilized *Scenedesmus quadricauda* was obtained under a semicontinuous mode of operation with hyperconcentrated cell systems (up to 3.3 g dry wt/l). Similarly, using the same urban effluent, Proulx and de la Noüe[84] studied batch and semicontinuous cultures of chitosan-immobilized filaments of the cyanobacterium *Phormidium* sp. It was shown that within a few hours it was possible to achieve virtually 100% removal of inorganic nitrogen (ammonium, nitrate, and nitrite) and 80% removal of phosphate.

The techniques for immobilizing microalgae are still in their infancy, and much research will have to be done before practicable systems are eventually available. Considering that additional and preliminary handling of the inoculum is required, it is likely that immobilized algal systems will be restricted, at least for the immediate future, to specific applications intended for the production of high-value products or for coimmobilization systems in which microalgae enhance the production of substances of interest (e.g., enzymes)[84] by bacterial partners.

At least in the near future, the use of microalgae immobilized in beads for large-scale biotreatment of huge volumes of diluted wastewaters such as agricultural drainage is unlikely to be easy because just maintaining the beads in suspension calls for a considerable input of energy via aeration or agitation devices. However, a search for new immobilization supports may lead to the discovery of matrices of a density appropriate to keep particles floating or in suspension with wind action alone. Another possibility would be to explore other immobilization systems whereby microalgae are not entrapped in beads, but are attached to fixed surfaces. In any event, practical use of immobilized microalgae in the foreseeable future would appear to be confined to intensive systems only.

C. ADVANTAGES AND DISADVANTAGES

The key advantages of closed systems stem from the ability to control the environment. The greater the degree to which the environment is controlled, the greater the degree to which one can take advantage of species-specific physiological capabilities of bioconcentration and biodegradation. Closure prevents contamination of the culture by undesirable species. Control of insolation, temperature, pH, and nutrient concentrations allows one to further optimize the environment to favor the performance of the chosen organism, and yields are generally more than double those achieved in open systems.

The major disadvantage to closed systems, however, is cost. Further development of closed systems will require more field testing, together with improved engineering and materials, in conjunction with advances in our ability to take advantage of species-specific metabolic attributes. The principal materials required in these systems are plastic containment devices and electronic monitoring and control components. Prices of plastic will closely parallel the cost of oil and thus are not easily controllable. Electronic monitoring and control components are continually improving in sophistication and decreasing in price. However, the most significant improvements in the cost/benefit ratio will probably derive from strain selection and genetic engineering — biotechnological methods which may enhance the physiological capabilities of individual cells by one or more orders of magnitude.[18,91] Not only will such specialized microorganisms require a closed system for their optimal growth and performance; they will also render it cost effective.

IV. HARVESTING

Since the pioneering work of the first investigators who proposed microalgal culture for the treatment of wastewaters,[7,13,24,92,93] harvesting the biomass produced has been the bottleneck for this type of technology due to the small size (<20 μm) of most species cultured.

Almost all conceivable means of harvesting algae have been attempted, involving methods which are either purely physical (sedimentation, centrifugation, sand filtration, flotation, microstraining, electroflocculation, ultrasonic treatment, and vacuum filtration), physicochemical (precipitation, pH treatment, ion exchange, chemical flocculation, and autoflocculation), or biological (bioflocculation and filtration by filter feeders). Several of these techniques were studied early on in the hope of finding one that would be economical and usable for large-scale operations.[94] Since then, efforts have been made to develop these and other suitable technologies.[95,96] It can be said that appropriate answers have been found from a technical point of view; there persists, however, incompatibility between methods which work efficiently and their excessive cost.[97] This situation has prompted some workers to bypass the harvesting problem by immobilizing the algae used for biotreatment processes,[59,81] but much research has yet to be done before even assessing the viability of this option for large-scale operation.

The most successful techniques, according to Mohn,[95] are centrifugation, filtration, and flocculation. Centrifugation, however, calls for an initial biomass concentration of ~10 g dry weight/l,[94] which is a condition impossible to meet in natural production systems. Therefore, even if centrifugation is to be chosen, it should be preceded by an additional preconcentration step. Chemical flocculation is a possible way of concentrating microalgae in preparation for an eventual flotation step.[65,96,98] This technique may, however, lead to secondary problems that can be worse than directly discharging the algae into receiving water bodies. For example, aluminum salts, which are an effective flocculant,[99] lead to acidification of the effluent and the presence of an undesirable or noxious metal.[100,101] Other flocculants, such as polyelectrolytes, have been banned in various countries because of their toxicity and/or carcinogenicity.[100] Reduction of the quantity of inorganic flocculants required is one way to partially alleviate these problems. This can be achieved through the use of organic polymers or ozone prior to flocculation with inorganic flocculants,[102] but these methods do not resolve all the difficulties, particularly the elevation of pH.

One flocculating agent of natural origin which does not produce many of the above problems is chitosan, a partially deacetylated *N*-acetyl glucosamine derivative of chitin. It is widely available and is effective for flocculating microalgae.[69,90,103] Although process optimization reduces the concentrations required for efficacy,[104] the price of chitosan (~$6/lb, U.S. 1987) precludes its use for large-scale applications. Moreover, the narrow pH interval (6.8 ± 0.2)[99] is difficult to attain for large volumes of algal culture, since actively photosynthesizing microalgae cause a considerable increase in pH (to pH 8 and often higher).

Some hope has emerged from a recent study on bioflocculation of microalgae which is accomplished by polymers which they themselves excrete — a phenomenon now known to occur at least among the cyanobacteria.[97] Bioflocculation does not require the addition of chemicals. This is in contrast to autoflocculation, which requires the presence of ions, especially calcium and phosphate,[105] at a minimal concentration of 40 mg/l (for Ca). Since natural freshwaters are generally below this concentration,[106] bioflocculation deserves careful study as a possible solution to the problem of harvesting.

Although bioflocculation appears unlikely with common species such as *Scenedesmus* spp.,[58] other photosynthetic microorganisms such as cyanobacteria have proved to be able to flocculate naturally and effectively.[107] The polymers excreted by some cyanobacteria, *Phormidium* J-1 for example,[108] appear to be promising in this regard. The polymeric sheath of another *Phormidium* species, *P. bohneri*,[109] efficiently traps debris or cells in cultures of this species which exhibit spontaneous formation of homogeneous flocs that rapidly settle (~30 cm/min). Indeed, the settling of this species is so rapid that efficient agitation of the medium is required to keep the flocs in suspension. This may preclude the use of such species for very large culture operations where, at best, paddle wheels appear to be a realistic mixing device, but may not provide sufficient mixing.

These features of bioflocculation, which have to be more fully understood and documented, may well justify the use of cyanobacteria rather than chlorophytes for biotreatment processes. The aggregation mechanism(s) of *Phormidium* appear(s) to be related to the hydrophobicity of its membrane proteins,[21] a phenomenon also occurring in *Dunaliella* spp., in which the process is salinity dependent.[110] Once again, a better understanding of the physiology of microalgae will lead to applications.

For various reasons, not the least of which is our imperfect knowledge of many facets of the behavior of cyanobacteria, the use of small chlorophytes will continue to be favored. One alternative to harvesting these small cells by physical or physicochemical means is to have the biomasses filtered out by invertebrates, i.e., to rely upon the aquatic food chains, which can even lead to fish as the final step. Such schemes have been employed, for the most part on an experimental level, using zooplankton, bivalves, and fish.[111-120] However, this alternative only partially solves the problem, since a significant concentration of nutrients and metabolic by-products remains in the final effluent. Obtaining an effluent devoid of these substances, therefore, still calls for a finishing step involving the incompletely solved problem of harvesting the final algal biomass.

V. BENEFICIAL BY-PRODUCTS

Economic processing of wastewaters is enhanced if marketable products are simultaneously biosynthesized with water treatment. In many cases, the value of the substances produced may be able to offset the high costs of mass culture and harvesting. Since the manufacture of specific secondary metabolites necessitates maintenance of cultures that are composed primarily of a single production species, economic considerations may favor the implementation of closed systems. Large-scale culture of single species of algae has been accomplished for frondose marine algae[9] and for several species of microalgae.[31,49,121] Many valuable substances can be produced from microalgae grown in wastewater, but the accumulation of toxic compounds from such environments does preclude the culture of algal food products such as *Porphyra* (nori), *Laminaria* (kombu), and *Chlorella*.[122]

A. COLORANTS

The gathering of light for photosynthesis is accomplished by a wide variety of plant pigments. Various species of algae often contain specialized or high concentrations of pigments as an adaptation to their particular environment; phycobiliprotein pigments are found in the Cyanobacteria, Cryptophyta, and Rhodophyta. These fluorescent red and blue pigments are used as food colorants,[123,124] fermentation additives,[125] and phycofluors.[126-128] A cyanobacterium, *Spirulina* sp., has been cultured in slightly brackish waters in 10-million-l volumes primarily as a protein supplement, but also for extracts of its blue phycocyanin pigment.[129] Several species of red algae are cultivated on floating marine farms in Asia. Extensive *Porphyra* farms place the algae on floating rope nets along bays and quiet coasts of Japan, Korea, and China.[122] These same farming techniques may be applied to yield red phycoerythrins from algae cultivated in agricultural drainage.

B. POLYSACCHARIDES

Soilborne species of *Chlamydomonas sajao* secrete large amounts of polysaccharides that condition soils by increasing wet aggregate stability and improve stability to slaking.[130,131] Polysaccharides secreted by *Porphyridium* spp. have been tested for enhanced recovery of oils which could replace xanthan or guar gums.[129] Long-chain polymers are essentially used to drag out fossil fuels, thereby reducing the residual oil saturation. Use is based on a ratio of approximately 500 g of polymer per stock tank barrel of oil.[132]

Phycocolloid-producing species, including *Gracilaria folifera*, *Hypnea musciformis*, *Chondrus crispus*, and *Neoagardhiella baileyi*, have been tested for nutrient scrubbing of marine systems containing sewage.[133] Several species of red algae contain commercially important monopolysaccharides such as agar, agarose, and carrageenan[134] which may, additionally, act to remove metals.

C. SEALANTS

In nature, high-salinity salterns are lined at their bases with extensive crusts of salt crystals. The low-salinity evaporating ponds in which crystallization occurs are sealed from ground leakage by thick layers of algae and bacteria which prevent the dissolution of the crystalline crust lining. Algal mats in these environments are dominated by cyanobacteria, but the species composition varies from location to location.[135] The algal mats improve salt recovery[136] by increasing the rate of evaporation as a result of absorption of solar irradiance and reduced input of new water.[137] Nutrient uptake by the algal mat and promotion of ecosystem balance improves the quality of the salt produced.[138] This same principle of bottom sealing by algal/bacterial mats has been applied to large (11,000-acre) inland agriculture evaporation systems, in which growth can continue without disruption of the bottom.[139]

D. VITAMINS AND OTHER PRODUCTS

The first industrial production of vitamins by algae was accomplished for β-carotene from *Dunaliella*.[28,29,140-144] Species of this alga grow in conditions of high salinity up to and including saturated concentrations of sodium chloride, in which they produce β-carotene as a protective pigment.[145] Accumulations of carotenoids in *Dunaliella salina* attain 13% by weight[146] and represent the highest cellular content of carotenoids found in nature. In dense cultures (~1 g dry weight/l), the uptake of ammonium can reach 1 mmol/d, with a doubling time of 1 d. Under these high-nutrient conditions, the alga normally contains less than 1% carotenoids.

Numerous other products, including biomass,[147] oils,[148,149] halogenated compounds,[150] and biologically active compounds,[151-154] have potential as by-products of agricultural wastewater treatment. Proof of industrial viability of these and other novel products is being awaited.

VI. CULTURE COLLECTIONS

Here we provide a list of groups and institutions from which one may obtain unialgal and axenic isolates of identified species of microalgae. In most cases, a catalog is available. The fees vary from supplier to supplier, but are generally quite modest. Of the algae companies listed, large quantities suitable for field inoculations are available from R & A Plant Soil, Inc. The catalog from the Microalgae Culture Collection (SERI) contains useful information about the characteristics of the algal species in field cultures. Many of the suppliers have traded species, especially with foreign collections. It is highly recommended that isolates be obtained from the closest source, since long shipment times will be deleterious to the health of the culture.

American Type Culture Collection
ATCC Catalogue of Protists — Algae and Protozoa
Order Department
12301 Parklawn Drive
Rockville, MD 20852
Telephone: (800) 638-6597
              (301) 881-2600

AquaOL, Inc.
P.O. Box 5565
Austin, TX 78763
Telephone: (512) 477-6332

Carolina Biological Supply Company
2700 York Road
Burlington, NC 27215
Telephone: (800) 547-1733
         (503) 656-1641

Center for the Culture of Marine Phytoplankton
Bigelow Laboratory for Ocean Sciences
McKown Point
West Boothbay Harbor, ME 04575
Telephone: (207) 633-2173

Chlamydomonas Genetics Center
Department of Botany
Duke University
Durham, NC 27706
Telephone: (919) 684-5243

Culture Center for Algae and Protozoa
36 Storey's Way
Cambridge CB3 ODT
U.K.
Telephone: Cambridge (0223) 61738

Culture Collection of Algae
Department of Botany
The University of Texas at Austin
Austin, TX 78712

Microalgae Culture Collection
Solar Energy Research Institute (SERI)
FTLB
1617 Cole Boulevard
Golden, CO 80401
Telephone: (303) 231-1842

The Microbial Culture Collection
The National Institute for Environmental Studies
Yatabe-cho
Tsukuba-gun
Ibaraki 305
Japan

R & A Plant Soil, Inc.
24 Pasco-Kahlotus Road
Pasco, WA 99301
Telephone: (509) 545-6867

Sammlung von Algenkulturen
Pflanzenphysiologisches Institut der Universität
Nikolausberger Weg 18
D-3400 Göttingen
Federal Republic of Germany

VII. SUMMARY

Algae possess a variety of physiological and metabolic adaptations which may be exploited to treat agricultural drainage.[18] The principal advantage to using algae is that they can utilize solar energy and inorganic nutrients (which are usually in plentiful supply in drainage waters) for growth, whereas most bacteria require supplemental growth substrates in the form of organic compounds. However, even under the best of conditions, algae grow more slowly than bacteria; thus, larger culture volumes may be required.

Open culture systems are likely to be most effective in limited circumstances where, for example, highly saline or alkaline influent waters are available; only under these specialized conditions will one be able to maintain a single species in culture. Open systems have two advantages: (1) they are relatively inexpensive to construct and (2) more than 40 years of research and practical experience can be drawn upon to guide successful operation. A major disadvantage of open culture systems is that they must be custom built to the environment; they often may not be suitable for the dependable culture of the increasingly broad variety of algal species or strains which are being selected or genetically manipulated for specific biodegradative or bioconcentrative abilities.

Closed systems are likely to become increasingly important in the treatment of wastewaters due to their potential for containment and control of many different algal species. They have already proved to be conducive to higher growth rates, by a factor of two or three, than comparable open systems. Strain selection and genetic manipulation have the potential to increase specified metabolic activities applicable to wastewater treatment by additional factors of up to one order of magnitude. As these advances occur and as more experience is gained with the variety of culture methodologies (concentrated suspensions, hyperconcentrated suspensions, and immobilization), the cost of constructing and maintaining closed system technologies is likely to converge with that of open systems.

Economically valuable by-products are a key factor worth considering in the choice of a treatment method. Such products, many with established markets, include colorants, polysaccharides, sealants, vitamins, oils, and biologically active compounds with potential pharmaceutical applications. Once again, the species-specific nature of the production of these substances will tend to favor further development and application of closed culture systems. In general, the use of algae for treatment of agricultural drainage is presently viable only under special circumstances, but will probably become increasingly tractable — and profitable — within the next decade.

REFERENCES

1. **Horne, A. J. and Nonomura, A. M.**, Drifting Macroalgae in Estuarine Water: Interactions with Salt Marsh and Human Communities, SERL Rep. No. 76-3, Sanitary Engineering Research Laboratory, University of California, Berkeley, 1976, 76.
2. **Bogan, R. H.**, Utilization of algae as an aid in sewage nutrient removals, in *Proc. 15th Industrial Waste Conf.*, Purdue University, Lafayette, IN, 1960, 68.
3. **Bogan, R. H., Albertson, O. E., and Pluntze, J. C.**, Use of algae in removing phosphorous from sewage, *J. Sanit. Eng. Div. Am. Soc. Civ. Eng.*, 86, 1, 1960.

4. **Oswald, W. J. and Golueke, C. G.**, Harvesting and processing of waste grown microalgae, in *Algae, Man and Environment*, Jackson, D. F., Ed., Syracuse University Press, Syracuse, NY, 1968, 371.
5. **Eliassen, R. and Tchobanoglous, G.**, Removal of nitrogen and phosphorous from waste water, *Environ. Sci. Technol.*, 3, 536, 1969.
6. **Dunstan, W. M. and Menzel, D. W.**, Continuous cultures of natural populations of phytoplankton in dilute, treated sewage effluent, *Limnol. Oceanogr.*, 16, 623, 1971.
7. **Oswald, W. J., Gotaas, H. B., Golueke, C. G., and Kellen, W. R.**, Algae in waste treatment, *Sewage Ind. Wastes*, 29, 437, 1957.
8. **Lincoln, E. P., Koopman, B., Bagnell, L. O., and Nordstedt, R. A.**, Aquatic system for fuel and feed production from livestock wastes, *J. Aquat. Eng. Res.*, 33, 159, 1986.
9. **LaPointe, B. E., Williams, L. D., Goldman, J. C., and Ryther, J. H.**, The mass culture of macroscopic marine algae, *Aquaculture*, 8, 9, 1976.
10. **Ryther, J. H., Goldman, J. C., Gifford, C. E., Huguenin, J. E., Wing, S., Clarner, J. P., Williams, L. D., and LaPointe, B. E.**, Physical models of integrated waste recycling-marine polyculture systems, *Aquaculture*, 5, 163, 1975.
11. **Round, F. E.**, *The Biology of the Algae*, 2nd ed., Edward Arnold, London, 1973, 278.
12. **Lobban, C. S. and Wynne, M. J., Eds.**, *The Biology of the Seaweeds*, Blackwell Scientific, Oxford, 1981, 786.
13. **Oswald, W. J., Gotaas, H. B., Ludwig, H. F., and Lynch, V.**, Algal symbiosis in oxidation ponds, *Sewage Ind. Wastes*, 25, 692, 1953.
14. **Burlew, J. S., Ed.**, *Algal Culture from Laboratory to Pilot Plant*, Carnegie Institute, Washington, D.C., 1953, 357.
15. **Gudin, C. and Chaumont, D.**, A biotechnology of photosynthetic cells based on the use of solar energy, *Biochem. Soc. Trans.*, 8, 481, 1980.
16. **Adey, W. H.**, The microcosm: a new tool for reef research, *Coral Reefs*, 1, 193, 1983.
17. **Doty, M. S.**, Status of marine agronomy, with special reference to the tropics, *Proc. Int. Seaweed Symp.*, 9, 35, 1978.
18. **Redalje, D. G., Duerr, E. O., de la Noüe, J., Mayzaud, P., Nonomura, A. M., and Cassin, R.**, Algae as ideal waste removers: biochemical pathways, in *Biotreatment of Agricultural Wastewater*, Huntley, M., Ed., CRC Press, Boca Raton, FL, 1989, chap. 7.
19. **Lobban, C. S., Harrison, P. J., and Duncan, M. J.**, *The Physiological Ecology of Seaweeds*, Cambridge University Press, London, 1985, 242.
20. **Round, F. E.**, *The Ecology of Algae*, Cambridge University Press, London, 1981, 653.
21. **Brown, R. L. and Beck, L. A.**, Subsurface agricultural drainage in California's San Joaquin Valley, in *Biotreatment of Agricultural Wastewater*, Huntley, M., Ed., CRC Press, Boca Raton, FL, 1989, chap. 1.
22. **Benemann, J. R., Tillett, D. M., and Weissman, J. C.**, Microalgae biotechnology, *Tibtech*, 5, 47, 1987.
23. **Cook, P. M.**, Some problems in the large-scale culture of *Chlorella*, in *The Culture of Algae*, Brunel, J., Prescott, G. W., and Tiffany, L. H., Eds., Charles F. Kettering Foundation, Yellow Springs, Ohio, 1950, 53.
24. **Gotaas, H. B. and Oswald, W. J.**, Utilization of solar energy for waste reclamation, in *Trans. 4th Conf. Use of Solar Energy — The Scientific Basis*, Carpenter, E. F., Ed., University of Arizona Press, Tucson, 1958, 95.
25. **Oswald, W. J.**, Current status of microalgae from wastes, *Chem. Eng. Prog. Symp. Ser.*, 65, 87, 1969.
26. **Oswald, W. J.**, Growth characteristics of microalgae cultured in domestic sewage: environmental effects on productivity, in *Prediction and Measurement of Photosynthetic Productivity*, Trebon Centre for Agricultural Publication and Documentation, Wageningen, The Netherlands, 1970.
27. **Oswald, W. J., Chen, P. H., Gerhardt, M. B., Green, F. B., Nurdogan, Y., Von Hippel, D. F., Newman, R. D., Shown, L., and Tam, C. S.**, The role of microalgae in removal of selenate from subsurface tile drainage, in *Biotreatment of Agricultural Wastewater*, Huntley, M., Ed., CRC Press, Boca Raton, FL, 1989, chap. 9.
28. **Ben-Amotz, A. and Avron, M.**, Glycerol, β-carotene and dry algal meal production by *Dunaliella*, in *Algae Biomass: Production and Use*, Shelef, G. and Soeder, C. J., Eds., Elsevier, Amsterdam, 1980, 603.
29. **Ben-Amotz, A. and Avron, M.**, Glycerol and β-carotene metabolism in the halotolerant alga *Dunaliella*: a model system for biosolar energy conversion, *Trends Biochem. Sci.*, 6, 297, 1980.
30. **Goldman, J. C.**, Physiological aspects in mass culture, in *Algae Biomass: Production and Use*, Shelef, G. and Soeder, C. J., Eds., Elsevier, Amsterdam, 1980, 343.
31. **Soong, P.**, Production and development of *Chlorella* and *Spirulina* in Taiwan, in *Algae Biomass: Production and Use*, Shelef, G. and Soeder, C. J., Eds., Elsevier, Amsterdam, 1980, 97.
32. **Zarrouk, G.**, Contribution à l'Étude d'une Cyanophyceae: Influence de Divers Facteurs Physiques et Chimiques sur la Croissance et la Photosynthèse de *Spirulina maxima* (Satch. et Gardner) Geitler, Thèse doctorat, Universite de Paris, Paris, 1977.

33. **Benemann, J. R., Koopman, B. L., Weissman, J. C., Eisenberg, D. M., and Oswald, W. J.,** Cultivation on Sewage of Microalgae Harvestable by Microstrainers, Final Report, Sanitary Engineering Research Laboratory, University of California, Berkeley, 1977.

34. **Clément, G.,** Production et constituants caracteristiques des algues *Spirulina platensis* et *maxima, Ann. Nutr. Aliment.*, 26, 477, 1975.

35. **Richmond, A. and Vonshak, A.,** Management of *Spirulina* mass culture, *Nova Hedwigia Z. Kryptoga-menkd.*, 83, 222, 1986.

36. **Howell, J. A., Frederickson, A. G., and Tsuchiya, H. M.,** Optimal and dynamic characteristics of a continuous photosynthetic algal gas exchanger, *Chem. Eng. Prog. Symp. Ser.*, 62, 56, 1966.

37. **Davis, E. A., Dedrick, J., French, C. S., Milner, H. W., Smith, J. H. C., and Spoehr, H. A.,** Laboratory experiments on *Chlorella* culture at the Carnegie Institution of Washington department of plant biology, in *Algal Culture from Laboratory to Pilot Plant*, Burlew, J. S., Ed., Carnegie Institute, Washington, D.C., 1953, 105.

38. **Miller, R. L., Frederickson, A. G., Brown, A. H., and Tsuchiya, H. M.,** Hydromechanical method to increase efficiency of algal photosynthesis, *Ind. Eng. Chem. Process Des. Dev.*, 3, 134, 1964.

39. **Oswald, W. J., Golueke, C. G., and Horning, D. O.,** Closed ecological systems, *J. Sanit. Eng. Div. Am. Soc. Civ. Eng.*, 91, 23, 1965.

40. **Shelef, G., Sabanas, M., and Oswald, W. J.,** An improved algatron reactor for photosynthetic life support systems, in *Proc. 14th Annu. Tech. Meet. Institution of Environmental Sciences*, Institution of Environmental Sciences, Mount Prospect, IL, 1968, 1.

41. **Emerson, R. and Arnold, W.,** A separation of the reactions in photosynthesis by means of intermittent light, *J. Gen. Physiol.*, 15, 391, 1932.

42. **Phillips, J. N. and Myers, J.,** Growth rate of *Chlorella* in flashing light, *Plant Physiol.*, 29, 152, 1954.

43. **Marra, J.,** Effect of short-term variations in light intensity on photosynthesis of marine phytoplankton: a laboratory simulation study, *Mar. Biol.*, 46, 203, 1978.

44. **Laws, E. A., Terry, K. L., Wickman, J., and Chalup, M. S.,** A simple algal production system designed to utilize the flashing light effect, *Biotechnol. Bioeng.*, 25, 2319, 1983.

45. **Laws, E. A.,** Use of the flashing light effect to stimulate production in algal mass cultures, *Nova Hedwigia Z. Kryptogamenkd.*, 83, 230, 1986.

46. **Goldman, J. C.,** Temperature effects on phytoplankton growth in continuous culture, *Limnol. Oceanogr.*, 22, 932, 1977.

47. **Dewey, D. R.,** System for Photosynthesis, U.S. Patent 2,732,663, 1956.

48. **Gudin, C.,** Method of Growing Plant Cells, U.S. Patent 3,955,317, 1976.

49. **Balloni, W., Materassi, R., Filpi, C., Sili, C., Vincenzini, M., Ena, M., and Florenzano, G.,** *Il Metodo di Trattamento a Batteri Fotosintetici delle Acque di Scarico*, Final Report, Centro di Studio dei Microorganismi Autotrofi, Grafiche Cappelli, Florence, 1982, 206.

50. **Hills, C. B.,** Method for Growing a Biomass in a Closed Tubular System, U.S. Patent 4,473,970, 1984.

51. **Huntley, M., Bardach, J., Booth, C., Bratkovich, A., Coughran, C., Jordan, J., Selover, D., Redalje, D., and Wahlberg, D.,** Design and Analysis of a Closed, Semi-Continuous Microalgae Culture Facility for the Purpose of Producing Fuels, Tech. Rep., Solar Energy Research Institute, Golden, CO, 1985, 139.

52. **Raymond, L. P.,** Mass Algal Culture System, U.S. Patent 4,320,594, 1982.

53. **McGriff, E. and McKinney, R.,** The removal of nutrients and organics by activated algae, *Water Res.*, 10, 1115, 1972.

54. **McKinney, R., McGriff, E., Sherwood, R. J., Wahbeh, N. V., and Newport, D. W.,** Ahead: activated algae?, *Water Waste Eng.*, 8, 51, 1971.

55. **Hendricks, F. and Bosman, J.,** The removal of nitrogen from an inorganic industrial effluent by means of intensive algal culture, *Prog. Water Technol.*, 12, 651, 1980.

56. **Doran, M. and Boyle, W.,** Phosphorous removal in activated algae, *Water Res.*, 13, 805, 1979.

57. **Humenick, F. and Hanna, G., Jr.,** Algal bacterial symbiosis for removal and conservation of wastewater nutrients, *J. Water Pollut. Control Fed.*, 43, 580, 1971.

58. **Lavole, A. and de la Noüe, J.,** Hyperconcentrated cultures of *Scenedesmus obliquus*. A new approach for wastewater biological tertiary treatment?, *Water Res.*, 19, 1437, 1985.

59. **Chevalier, P. and de la Noüe, J.,** Efficiency of immobilized hyperconcentrated algae for ammonium and orthophosphate removal from wastewaters, *Biotechnol. Lett.*, 7, 395, 1985.

60. **Lewin, J. C. and Lewin, R. A.,** Autotrophy and heterotrophy in marine and littoral diatoms, *Can. J. Microbiol.*, 6, 127, 1960.

61. **Hellebust, J. A. and Lewin, J.,** Heterotrophic nutrition, in *Biology of Diatoms*, Werner, D., Ed., University of California Press, Berkeley, 1977, 169.

62. **Bollman, R. C. and Robinson, G. C. C.,** The kinetics of organic acid uptake by three chlorophyta in axenic culture, *J. Phycol.*, 13, 1, 1977.

63. **Kawaguchi, K.,** Microalgae production in Asia, in *Algae Biomass: Production and Use*, Shelef, G. and Soeder, C. J., Eds., Elsevier, Amsterdam, 1980, 25.

64. **Vincent, W. F. and Goldman, C. R.,** Evidence for algal heterotrophy in Lake Tahoe, California, *Limnol. Oceanogr.*, 25, 89, 1980.
65. **Becker, E. W. and Venkataraman, L. V.,** Biotechnology and exploitation of algae, the Indian approach, *Deutsche Ges. Tech. Zusammenarbeit*, 216, 1982.
66. **Kreuzberg, K., Reznicezk, G., and Klöck, G.,** Properties of algal biomass production and the parameters determining its fermentative degradation, paper presented at the 3rd Conf. Energy from Biomass, Venice, Italy, March 25 to 29, 1985.
67. **Kaplan, D., Richmond, A. E., Dubinsky, Z., and Aaronson, S.,** Algal nutrition, in *Handbook of Microalgal Mass Culture*, Richmond, A., Ed., CRC Press, Boca Raton, FL, 1986, 147.
68. **Neilson, A. H., Blankley, W. F., and Lewin, R. A.,** Growth with organic carbon and energy sources, in *Handbook of Phycological Methods: Culture and Growth Measurements*, Stein, J., Ed., Cambridge University Press, London, 1973, chap. 18.
69. **Lalucat, J., Imperial, J., and Parés, R.,** Utilization of matter in *Chlorella* sp. VJ79, *Biotechnol. Bioeng.*, 26, 677, 1984.
70. **Robinson, A. L., Mak, A., and Trevan, M. D.,** Immobilized algae: a review, *Process Biochem.*, 21, 122, 1986.
71. **Mattiasson, B.,** Immobilized viable cells, in *Immobilized Cells and Organelles*, Mattiasson, B., Ed., CRC Press, Boca Raton, FL, 1983, chap. 2.
72. **Baillez, E., Largeau, C., Berkaloff, C., and Casadevall, E.,** Immobilization of *Botryococcus braunii* in alginate: influence on chlorophyll content, photosynthetic activity and degeneration during batch cultures, *Appl. Microbiol. Biotechnol.*, 23, 361, 1986.
73. **Brodelius, P. and Mosbach, K.,** Immobilized plant cells, *Adv. Appl. Microbiol.*, 28, 1, 1982.
74. **Hahn-Hägerdal, B.,** Co-immobilization involving cells, organelles and enzymes, in *Immobilized Cells and Organelles*, Mattiasson, B., Ed., CRC Press, Boca Raton, FL, 1983, chap. 5.
75. **Papageorgiu, G. C. and Lagoyanni, T.,** Immobilization of photosynthetically active cyanobacteria in glutaraldehyde cross-linked albumin matrix, *Appl. Microbiol. Biotechnol.*, 23, 417, 1986.
76. **Enfors, S. O. and Mattiasson, B.,** Oxygenation of processes involving immobilized cells, in *Immobilized Cells and Organelles*, Mattiasson, B., Ed., CRC Press, Boca Raton, FL, 1983, chap. 3.
77. **Chevalier, P.,** Etude du Comportement de Microalgues et de Bactéries Immobilisées dans un Gel de Carraghenine: Croissance et Cinétique de Production d'alpha-Amylase, Ph.D. thesis, Université Laval, Québec, Canada, 1987.
78. **Chevalier, P. and de la Noüe, J.,** Enhancement of alpha-amylase production by immobilized *Bacillus subtilis* in an airlift fermenter, *Enzyme Microb. Technol.*, 9, 53, 1987.
79. **Kuu, W. Y. and Polack, J. A.,** Improving immobilized biocatalysts by gel phase polymerization, *Biotechnol. Bioeng.*, 25, 1995, 1983.
80. **Wickstrom, P., Swajcer, E., Brodelius, K., Nilsson, K., and Mosbach, K.,** Formation of alpha-keto acids from amino acids using immobilized bacteria and algae, *Biotechnol. Lett.*, 4, 153, 1982.
81. **Chevalier, P. and de la Noüe, J.,** Wastewater nutrient removal with microalgae immobilized in carrageenan, *Enzyme Microb. Technol.*, 7, 621, 1985.
82. **Klein, J.,** Functions of polysaccharides in biotechnology, in *Biotechnology of Marine Polysaccharides*, Colwell, R., Pariser, E. R., and Sinskey, A. J., Eds., *Proc. 3rd Annu. MIT Sea Grant College Program Lectures and Seminar*, McGraw-Hill, New York, 1985, 3.
83. **Chibata, I. and Tosa, T.,** Transformation of organic compounds by immobilized microbial cells, *Adv. Appl. Microbiol.*, 22, 1, 1977.
84. **Proulx, D. and de la Noüe, J.,** Removal of macronutrients from wastewaters by immobilized microalgae, in *Proc. Int. Symp. Immobilized Enzymes and Cells*, Moo-Young, M., Ed., Elsevier, New York, 1988, 301.
85. **Gudin, C. and Thomas, D.,** Production de polysaccharides sulfatés par un photobioréacteur à cellules immobilisées de *Porphyridium cruentum*, *C. R. Acad. Sci. Paris Ser. III*, 293, 35, 1981.
86. **Baillez, E., Largeau, C., Casadevall, E., and Berkaloff, C.,** Effets de l'immobilisation en gel d'alginate sur l'algue *Botryococcus braunii*, *C. R. Acad. Sci. Paris Ser. III*, 296, 199, 1983.
87. **Rao, K. K. and Hall, D. O.,** Photosynthetic production of fuels and chemicals in immobilized systems, *Trends Biotechnol.*, 2, 124, 1984.
88. **Muallem, A., Bruce, D., and Hall, D. O.,** Photoproduction of H_2 and $NADPH_2$ by polyurethane-immobilized cyanobacteria, *Biotechnol. Lett.*, 5, 365, 1983.
89. **Adlercreutz, P. and Mattiasson, B.,** Oxygen supply to immobilized cells. I. Oxygen production by immobilized *Chlorella pyrenoidosa*, *Enzyme Microb. Technol.*, 4, 332, 1982.
90. **Musgrave, S. C., Kerby, N. W., Codd, G. A., and Stewart, W. D. P.,** Sustained ammonia production by immobilized filaments of the nitrogen-fixing cyanobacterium *Anabaena* ATCC 27893, *Biotechnol. Lett.*, 4, 647, 1982.
91. **Sayler, G. S. and Blackburn, J. W.,** Modern biological methods: the role of biotechnology, in *Biotreatment of Agricultural Wastewater*, Huntley, M., Ed., CRC Press, Boca Raton, FL, 1989, chap. 5.

92. **Caldwell, D. H.**, Sewage oxidation ponds — performance, operation and design, *Sewage Works J.*, 18, 433, 1946.

93. **Oswald, W. J. and Gotaas, H. B.**, Photosynthesis in sewage treatment, *Trans. Am. Soc. Civ. Eng.*, 122, 73, 1957.

94. **Golueke, C. G. and Oswald, W. J.**, Harvesting and processing sewage-grown plankton algae, *J. Water Pollut. Control Fed.*, 37, 471, 1965.

95. **Mohn, H.**, Experiences and strategies in the recovery of biomass from mass cultures of microalgae, in *Algae Biomass: Production and Use*, Shelef, G. and Soeder, C. J., Eds., Elsevier, Amsterdam, 1980, 547.

96. **Richmond, A. and Becker, E. W.**, Technological aspects of mass cultivation — a general outline, in *Handbook of Microalgal Mass Culture*, Richmond, A., Ed., CRC Press, Boca Raton, FL, 1986, 245.

97. **Benemann, J. R., Koopman, B. L., Weissman, J. C., Eisenberg, D. M., and Goebel, R.**, Development of microalgae harvesting and high-rate pond technologies in California, in *Algae Biomass: Production and Use*, Shelef, G. and Soeder, C. J., Eds., Elsevier, Amsterdam, 1980, 457.

98. **De Pauw, N. and Van Vaerenbergh, E.**, Microalgal wastewater treatment systems: potentials and limits, in *Phytoepuration and the Employment of the Biomass Produced*, Ghetti, P. F., Ed., Centro Ricerche Produzione Animali, Reggio Emilia, Italy, 1983, 211.

99. **Lavoie, A., de la Noüe, J., and Sérodes, J. B.**, Récupération de microalgues en eaux usées: étude comparative de divers agents floculants, *Can. J. Civ. Eng.*, 11, 266, 1986.

100. **Dodd, J. C.**, Algae production and harvesting from animal wastewaters, *Agric. Wastes*, 1, 23, 1979.

101. **Shelef, G., Oron, G., and Moraine, R.**, Economic aspects of microalgae production on sewage, *Arch. Hydrobiol. Beih. Ergeb. Limnol.*, 11, 281, 1978.

102. **Shelef, G., Sukenik, A., and Green, M.**, Separation and harvesting of marine microalgal biomass, *Nova Hedwigia Z. Kryptogamenkd.*, 83, 245, 1986.

103. **Nigam, B. P., Ramanathan, P. K., and Venkataraman, L. V.**, Application of chitosan as a flocculant for the cultures of the green alga *Scenedesmus acutus*, *Arch. Hydrobiol.*, 88, 378, 1980.

104. **Morales, J., de la Noüe, J., and Picard, G.**, Harvesting marine microalgae species by chitosan flocculation, *Aquaculture Eng.*, 4, 257, 1985.

105. **Sukenik, A. and Shelef, G.**, Algal autoflocculation — verification and proposed mechanism, *Biotechnol. Bioeng.*, 26, 142, 1984.

106. **Reid, G. K. and Wood, R. D.**, *Ecology of Inland Waters and Estuaries*, 2nd ed., D Van Nostrand, New York, 1976.

107. **Talbot, P., Dauta, A., and de la Noüe, J.**, Etude d'une algue filamenteuse pour l'épuration d'effluents, aspects fondamentaux: caractéristiques de l'algue *Phormidium bohneri*, poster presented at Association Canadienne pour la Recherche sur la Pollution de l'Eau et sa maîtrise, Québec, October 30, 1986.

108. **Fatton, A. and Shilo, M.**, *Phormidium* J-1 bioflocculant: production and activity, *Arch. Microbiol.*, 139, 421, 1984.

109. **Talbot, P. and de la Noüe, J.**, Evaluation of *Phormidium bohneri* for solar biotechnology, in *Algal Biotechnology*, Stadler, T., Mollion, G., Verdus, M. C., Karamanos, Y., Morvan, H., and Christiaen, D., Eds., Elsevier, Amsterdam, 1988, 403.

110. **Curtain, C. C. and Snook, H.**, Method for Harvesting Algae, PCT/AV82/00165, October 8, 1982; U.S. Serial No. 511 135, June 7, 1983.

111. **Dinges, R.**, The availability of *Daphnia* for water quality improvement and as an animal food source, in Proc. Wastewater Use in the Production of Food and Fiber, EPA 660/2-74-041, U.S. Environmental Protection Agency, Washington, D.C., 1974, 142.

112. **Goldman, J. C. and Ryther, J. H.**, Waste reclamation in an integrated food chain system, in *Biological Control of Water Pollution*, Tourbier, J. and Pierson, R. W., Eds., University of Pennsylvania Press, Philadelphia, 1976, 197.

113. **De Pauw, N., Verlet, H, and De Leenheer, L.**, Heated and unheated outdoor cultures of marine algae with animal manure, in *Algae Biomass: Production and Use*, Shelef, G. and Soeder, C. J., Eds., Elsevier, Amsterdam, 1980, 315.

114. **De Pauw, N., De Leenheer, L., Laureys, P., Morales, J., and Reartes, J.**, Cultures d'algues et d'invertébrés sur déchets agricoles, in *La Pisciculture en Etang*, Billard, R., Ed., I.N.R.A., Paris, 1980, 189.

115. **Pieters, A. J. H. and Le Roux, J.**, Algal culture using wastewater, in 5th Symp. Aquaculture in Wastewater, CSIR, Pretoria, South Africa, November 1980.

116. **Edwards, P.**, Review of recycling wastes into fish with emphasis on the tropics, *Aquaculture*, 21, 261, 1980.

117. **Groeneweg, J. and Schlütter, M.**, Mass production of freshwater rotifers on liquid wastes. II. Mass production of *Brachionus rubens* Ehrenberg 1938 in the effluent of high-rate algal ponds for the treatment of piggery waste, *Aquaculture*, 25, 25, 1981.

118. **Tarifeno-Silva, E., Kawasaki, L. Y., Yu, D. P., Gordon, M. S., and Chapman, D. J.**, Aquacultural approaches to recycling of dissolved nutrients in secondarily treated domestic wastewaters. II. Biological productivity of artificial food chains, *Water Res.*, 19, 51, 1982.

119. **Proulx, D. and de la Noüe, J.**, Harvesting *Daphnia magna* grown on urban tertiarily-treated effluents, *Water Res.*, 19, 1319, 1985.
120. **Proulx, D. and de la Noüe, J.**, Harvesting *Daphnia magna* grown on urban wastewaters tertiarily-treated with *Scenedesmus* sp., *Aquacult. Eng.*, 4, 93, 1985.
121. **Barclay, W. and McIntosh, R. P., Eds.**, Algal Biomass Technologies, *Nova Hedwigia Z. Kryptogamenk.*, Vol. 83, 1986.
122. **Nonomura, A. M., Woessner, J., and West, J. A.**, Making Seaweeds Worth Eating, Carolina Supply Co., Burlington, NC, 1980.
123. **Langston, M. and Mainz, I.**, Acid-Soluble *Spirulina*-Blue Colorant Preparation by Treating *Spirulina*-Blue with Protease and Alkaline Medium and Acidifying, U.S. Patent 4,400,400, 1983.
124. **Ogawa, K., Tezuka, S., Tsucha, Y., Tanabe, Y., and Iwamoto, H.**, Blue Chewing Gum Colored by Phycocyanin Obtained by Extraction of *Spirulina*, Japanese Patent 79138156, 1979.
125. **Tanabe, Y. and Iwamoto, H.**, Phycocyanin for Fermented Milk Production, Japanese Patent 79095770, 1979.
126. **Glazer, A. J. and Stryer, L.**, Phycofluor probes, *Trends Biochem. Sci.*, 9, 423, 1984.
127. **Ong, L. J., Glazer, A. J., and Waterbury, J. B.**, An unusual phycoerythrin pigment from a marine cyanobacterium, *Science*, 224, 80, 1984.
128. **Stryer, L. and Glazer, A. J.**, Novel Phycobiliprotein Fluorescent Conjugates, U.S. Patent Application 483,006, 1985.
129. **Richmond, A. E.**, Microalgalculture, in *Handbook of Microalgal Mass Culture*, Richmond, A., Ed., CRC Press, Boca Raton, FL, 1986, 369.
130. **Metting, B.**, Population dynamics of *Chlamydomonas sajao* and its influence on soil aggregate stabilization in the field, *Appl. Environ. Microbiol.*, 51, 1161, 1986.
131. **Metting, B.**, Dynamics of wet and dry aggregate stability from a three-year microalgal soil conditioning experiment in the field, *Soil Sci.*, 143, 139, 1987.
132. **Crump, J. R., Claridge, E. L., and Young, R. D.**, Chemicals for Chemical Flooding in Enhanced Oil Recovery, DOE/ET/10145-66, National Technical Information Service, U.S. Department of Commerce, Springfield, VA 1981.
133. **Hansen, J. A., Packard, J. E., and Doyle, W. T.**, Mariculture of Red Seaweeds, Rep. #T-CSGCP-002, California Sea-Grant College Program, University of California, San Diego, 1981.
134. **Renn, D. W.**, Marine algae and their role in biotechnology, in *Biotechnology in the Marine Sciences*, Colwell, R., Pariser, E., and Sinskey, A., Eds., John Wiley & Sons, New York, 1984, 191.
135. **Davis, J. and Lipkin, Y.**, *Lamprothamnium* prosperity in permanently hypersaline water, *Schweiz. Z. Hydrol.*, 48, 240, 1986.
136. **Nonomura, A. M.**, Brine Technology in Salt Ponds, CPR/85/066/A/01/99, United Nations Development Program, Beijing, 1986.
137. **Davis, J. S.**, Biological communities of a nutrient enriched salina, *Aquat. Bot.*, 4, 23, 1978.
138. **Davis, J. S.**, Experiences with *Artemia* at solar saltworks, in *The Brine Shrimp Artemia*, Vol. 3, Persoone, G., Sorgeloos, P., Roels, O., and Jaspers, E., Eds., Universa Press, Wetteren, Belgium, 1980, 52.
139. **Davis, J. S.**, personal communication.
140. **Avron, M. and Ben-Amotz, A.**, Alga Strain, U.S. Plant Patent 4511, 1980.
141. **Ben-Amotz, A. and Avron, M.**, On the factors which determine massive β-carotene accumulation in the halotolerant alga *Dunaliella bardawil*, *Plant Physiol.*, 72, 593, 1983.
142. **Ben-Amotz, A. and Avron, M.**, Accumulation of metabolites by halotolerant algae and its industrial potential, *Annu. Rev. Microbiol.*, 37, 95, 1983.
143. **Ben-Amotz, A., Katz, A., and Avron, M.**, Accumulation of β-carotene in halotolerant algae: purification and characterization of β-carotene-rich globules from *Dunaliella bardawil* (Chlorophyceae), *J. Phycol.*, 18, 529, 1982.
144. **Nonomura, A. M.**, Process for Producing a Naturally-Derived Carotene/Oil Composition by Extraction from Algae, U.S. Patent 4,680,314, 1985.
145. **Ben-Amotz, A.**, Beta-carotene enhancement and its role in protecting *Dunaliella bardawil* against injury by high irradiance, *Nova Hedwigia Z. Kryptogamenkd.*, 8, 1323, 1986.
146. **Aasen, A. J., Eimhjellen, K. E., and Liaaen-Jensen, S.**, An extreme source of β-carotene, *Acta Chem. Scand.*, 23, 2455, 1969.
147. **Shelef, G. and Soeder, C. J., Eds.**, *Algae Biomass: Production and Use*, Elsevier, Amsterdam, 1980, 735.
148. **Wolf, F. R.**, *Botryococcus braunii:* an unusual hydrocarbon-producing alga, *Appl. Biochem. Biotechnol.*, 8, 249, 1983.
149. **Wolf, F. R., Nonomura, A. M., and Bassham, J. A.**, Growth and branched hydrocarbon production in a strain of *Botryococcus braunii* (Chlorophyta), *J. Phycol.*, 21, 388, 1985.
150. **Howard, B. M., Nonomura, A. M., and Fenical, W.**, Chemotaxonomy in marine algae. II. Secondary metabolite synthesis by *Laurencia nipponica* (Rhodophyta, Ceramiales) in unialgal culture, *J. Exp. Biochem. Ecol.*, 8, 329, 1980.

151. **Metting, B. and Pyne, J. W.,** Biologically active compounds from microalgae, *Enzyme Microbiol. Technol.,* 8, 386, 1986.
152. **Jacobsen, N. and Pedersen, E. K.,** Synthesis and insecticidal properties of derivatives of propane-1,3-dithiol (analogues of the insecticidal derivatives of dithiolane and trithiane from the alga *Chara globularis* Thuiller), *Pestic. Sci.,* 14, 90, 1983.
153. **Kao, C. Y. and Walker, S. E.,** Active groups of saxitoxin and tetrodotoxin as deduced from actions of saxitoxin analogues on frog muscle and squid axon, *J. Physiol.,* 323, 619, 1982.
154. **Rich, R.,** Chlorococcales, Tetrasporales inhibit mosquito larvae, *Phycol. Soc. Am. Newsl.,* 21, 10, 1985.
155. **Tchobanoglous, G. and Schroeder, E.,** *Water Quality: Characteristics, Modeling, Modification,* Addison-Wesley, Reading, MA, 1985, 768.

Chapter 9

THE ROLE OF MICROALGAE IN REMOVAL OF SELENATE FROM SUBSURFACE TILE DRAINAGE

William J. Oswald, Paris H. Chen, Matthew B. Gerhardt, F. Bailey Green,
Yakup Nurdogan, David F. Von Hippel, Robert D. Newman, Leslie Shown,
and Christine S. Tam

TABLE OF CONTENTS

I. INTRODUCTION

Selenium in various forms is present in subsurface tile drainage from agricultural lands in the western San Joaquin Valley of California, usually at concentrations ranging from 100 to 1400 µg/l. California has recommended an interim maximum mean monthly level of 5 µg/l for selenium in the San Joaquin River at Hills Ferry and downstream and 2 µg/l in wetlands.[1] Accordingly, selenium in excess of these levels may have to be removed from subsurface tile drainage waters prior to the legal disposal of these waters at specific locations.

Selenium in its oxidized form, selenate (SeO_4^{2-}), is highly soluble, whereas metallic selenium (Se^0) and the reduced forms, such as selenite (SeO_3^{2-}) and selenide (Se^{2-}), tend to be insoluble or to form insoluble precipitates with metals such as iron (Fe^{2+}). A four-step process thus seems to be indicated to remove selenium from water: (1) reduction of selenate to selenite, metallic selenium, and selenide; (2) precipitation of the reduced material with metal salts; (3) removal of the precipitate from the residual water; and (4) legal disposal of the precipitate. From a practical standpoint, this whole series of steps should have a net cost less than the guideline amount of U.S. $100/acre-ft (1987) of drainage water,[2] which is the estimated affordable cost based on the average net farm income per acre in the San Joaquin Valley.

Before discussing the role of algae in applying these four steps, we should mention the dearth of direct uptake of selenium by microalgal cultures. Selenium is a trace nutrient for microalgae[3,4] and, as such, there is some direct uptake.[5,6] While some algae are reported to contain relatively large amounts of selenium after growing in selenium-spiked waters,[7] the kind of "algae weeds" that grow naturally in drainage water show only a small uptake. Arthur[8] has analyzed dried algal sludges remaining from growth experiments on selenium-rich drainage water at Firebaugh, CA and found 5 mg/kg ash-free dry weight (AFDW).[9] This falls far short of the uptake needed to remove selenium concentrations of hundreds of micrograms per liter. For example, to obtain 1 kg of algae from a growing algal concentration of 250 mg/l, one must process 4000 l of water. If this water contained 300 µg/l of selenium, the amount of selenium involved would be 1.2 g of soluble selenium per kilogram (AFDW) of algae produced. To remove all selenium by direct uptake, the algae would need to contain 0.12% selenium rather than the 0.0005% selenium observed by Arthur. Thus, algae growing in drainage water would need to take up 240 times as much selenium as they did in earlier studies — an unlikely occurrence.

We have recently explored direct uptake by several algal species, since algae have been shown to bioaccumulate some elements.[10] As shown in Table 1, we found some direct uptake based on the residual selenium in the drainage water-based medium in which the algae were grown. Most of the selenium was probably in the form of selenate. Although some uptake is indicated, it is not sufficient to meet standards. Accordingly, we have sought methods other than direct uptake for the kind of "algae weeds" that grow in open drainage high-rate ponds (HRPs), which are shallow, continuous channel ponds mixed with paddle wheels.

A second point of importance is that algae growing in an open HRP release dissolved oxygen in amounts equal to at least 1.5 times their AFDW, making it likely that all selenium present in an HRP system will be in an oxidized form (SeO_4^{2-}) and, hence, too soluble to be incorporated in the precipitates that form profusely in the algal cultures growing on San Joaquin Valley drainage waters. There is thus little possibility that selenium can be precipitated and incorporated in the aerobic sludges that are removable by simple sedimentation.

II. ALGAL-BACTERIAL SELENIUM REMOVAL PROCESS

A. PHYSICAL LAYOUT

With these preliminary concepts in mind, we may now consider the relationship of microalgae to the four-step process outlined previously. The overall process of the algal-

TABLE 1
Selenium Removal from Agricultural Drainage Water by Algae

Algal culture	Residence time (d)	Soluble selenium (μg/l)		Selenium removal (%)
		Initial	Final	
High-rate pond	6	225	188	16
(mixed culture)	21	334	312	7
Dunaliella spp.	6	225	195	13
	21	334	256	23

bacterial selenium removal process we are exploring at the University of California, Berkeley, under the sponsorship of the California Department of Water Resources (DWR) and the Drainage Program of the U.S. Bureau of Reclamation (BOR), is illustrated schematically in Figure 1. As indicated in the figure, drainage water is first subjected to algal growth in an HRP to incorporate carbon from CO_2 and HCO_3^- and nitrogen from NO_3^- into algal cell biomass to provide organic matter for methane fermentation. Because direct removal of sulfate has proved costly, this biomass is concentrated by natural subsidence and then by gravity thickening followed by centrifugation to achieve a low sulfate-to-carbon ratio. The thickened material is diluted with recycled sludge supernatant and then introduced to the methane fermentation system (anaerobic digester) to produce methane for electrical power generation. This electrical power is to be used to operate the system, and there should be a surplus for other uses. In a full-scale system, carbon dioxide from the power generator will be recycled to maintain an HRP pH below 8.3 and to provide carbon needed for microalgal growth. Sludges from the methane fermentation system are to be introduced into a reduction chamber, where drainage water from the HRP will for approximately 24 h be in contact with the intensely anoxic environment created by algal sludges decomposing in the absence of oxygen. Here it is hoped that selenium will be reduced to one of its insoluble forms and will either become incorporated in bottom solids or remain suspended for final removal in a dissolved air flotation (DAF) unit. DAF unit sludges may be recycled to the reduction chamber or dewatered and dried on an underdrained sand bed for final disposal. The intent is to economically and dependably produce a clear water with selenium levels less than 5 μg/l. The extent to which this can be achieved will be determined when the full bench-scale system is in operation and analytical work can be completed on representative samples.

B. REDUCTION OF SELENATE

According to McKeown and Marinas,[11] the reduction of selenate to selenite is a slow process requiring a number of hours for its completion, even under the most ideal conditions. An important function of a reduction system is to convert soluble selenate to insoluble forms as quickly as possible to minimize reactor size. The reducing material should be sufficient in amount, low in cost, and dependable in supply. The extent to which fermented algae can meet these criteria is discussed below.

C. ALGAE AS A REDUCTANT

Concentrated waste-grown microalgal sludge originating from Richmond, CA sewage and produced during methane fermentation has been demonstrated to be a satisfactory reducing material, since it converts soluble selenate in drainage waters to insoluble forms. As shown in Figure 2, removal in 1:1 proportions and 12 h contact time exceeds 95% when the influent Se concentration is near 250 μg/l. To date, we have been unable to demonstrate such high degrees of removal using naturally settled fermented concentrates of algae grown in San Joaquin Valley drainage waters. Poor removal compared with Richmond algae is

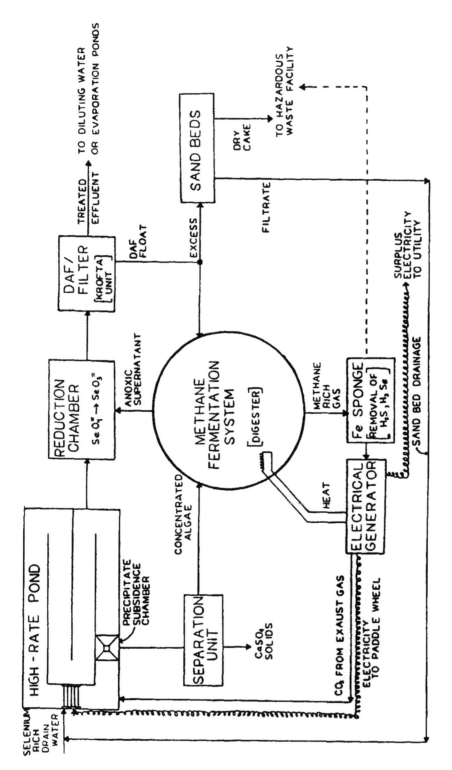

FIGURE 1.   Schematic diagram of proposed selenium removal plant currently under study by the Algae Research Group, University of California, Berkeley.

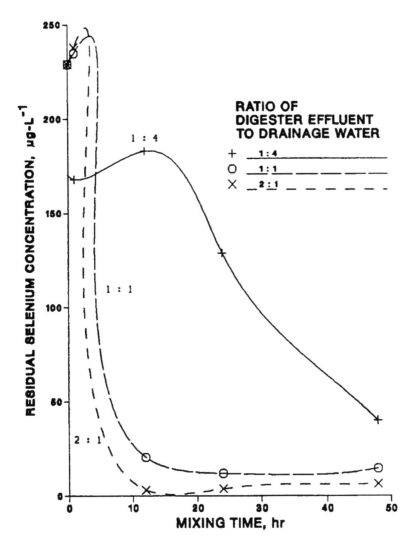

FIGURE 2. Micrograms selenium in effluent per liter of influent drain water as a function of contact time and digester mixture/drain water ratio.

believed to be caused by the presence of large amounts of dissolved and possibly insoluble sulfate in settled concentrates. As shown in Table 2, drainage water contains sulfate and nitrate in concentrations greater than that of carbon. Other chemical constituents of drainage water shown in Table 2 could also interfere with the reduction process. At existing levels, sulfate and nitrate will interfere with methane fermentation in an anaerobic environment. This interference can be avoided by decreasing the concentrations of sulfate and nitrate in the digester feed. This can be accomplished by dewatering the algal sludges and subsequently resuspending them in recycled digester supernatant in which only reduced forms of sulfur and nitrogen are present. We are able to gravity thicken the algal sludges to 2 to 3% total solids (TS), so 97 to 98% is water containing dissolved sulfate and nitrate. More of this water is removed by centrifugation or DAF and replaced with sulfate-free supernatant from the digester so that the methane fermentation might be more successful. We are currently studying the efficacy of improving methane fermentation by increasing the carbon-to-sulfur ratio.

As noted above, by comparing concentrations of 250 mg/l for algae and 3000 mg/l for sulfate, it seems very apparent that the amount of algae produced daily will always be

TABLE 2
Range of Individual Constituents Reported in San Joaquin
Valley Agriculture Drainage Waters[12]

Parameters	Concentration		
	Maximum	Minimum	Average
Calcium	590	55	369
Magnesium	583	54	161
Sodium	5,250	400	1,093
Potassium	7.0	0.9	4.0
Bicarbonate	391	132	214
Sulfate	10,100	607	2,282
Nitrate (as NO_3)	234	3.2	97
Chloride	2,680	110	987
Manganese	0.36	0.04	—[a]
Iron	12	0.04	2.1
Strontium	4.8	3.6	—[a]
Boron	34	1.4	12
Silica	60	7.2	26
Phosphates (as P)	0.15	0.01	0.04
Total organic carbon	14	1.0	4.6
pH	8.0	7.0	7.3
Temperature (°C)	21	7	16
Suspended solids	715	1	54
Total dissolved solids	20,700	1,930	5,342
Electrical conductivity (μmho/cm)	21,500	2,250	6,462

Note: Concentrations of chemical constituents are expressed as milligrams per
liter (mg/l) of constituent unless stated otherwise. Averages are based on
5 to 200+ individual samples.

[a]   Only two samples.

insufficient to reduce all of the sulfate in drainage water. Evidence from our early studies
(illustrated in Figure 2), however, indicates that reduction of all sulfate is not required to
precipitate selenium. Apparently, an insoluble precipitate is achieved at a redox potential
well above that of sulfate reduction. In any event, we intend to retain all reduced sludges
resulting from methane fermentation of the algae and a portion of that separated from the
final effluent. They will be retained within a lined reduction chamber with a hydraulic
residence time for the full volume of drainage flow of about 1 d. In practice, the reduction
chamber would be comprised of one or more deep anoxic ponds with covers to prevent
escape of odors and noxious gases and to capture combustible gases. The intensity of anoxia
in the reduction chambers may be controllable by varying the amount of digested algal
sludge applied to them.

D. COST OF ALGAE AS A REDUCTANT

The cost of microalgal sludge as a reductant will mainly depend on the cost of ponds
and the productivity of the algae. Our springtime productivity in the algal bench-scale
experimental production pond, using a 5-d residence time and a 20-cm depth, was about
100 kg/ha/d. This was increased during the summer by using a slightly greater depth (26
cm) and a shorter (3-d) residence time. A residence time as low as 3 d and a depth as great
as 30 cm seem to be feasible from May through August due to the higher growth rates
attainable in the summer. This is also the heaviest irrigation season and, thus, the time when
the greatest amount of drainage water will be produced. Within a depth of 30 cm and a
residence time of 3 d, productivity would approach 250 kg/ha/d. Higher productivities than

this are not likely, and 200 kg/ha/d is a more attainable goal.[13] Aside from the cost of ponds, which could be amortized over many years (possibly 50), there are the costs of carbon, phosphorus, and iron, which must be added to the drainage water to support algal growth. Carbon will be recycled as CO_2 from the combustion of methane. According to our current studies, purchased phosphorus will be required at the rate of about 2.5 kg/ha/d. The cost of this will be about U.S. $5.00/ha/d (1987). Iron will be added as Fe-EDTA initially, but natural polymers and stabilizers produced by pond organisms have been found to eliminate the need for continuous application of EDTA. Ferric or ferrous chloride will be used in polishing the final effluent; hence, a low-cost source of iron will be available at the site. Certainly the cost of iron addition will not exceed U.S. $5.00/ha/d (1987).

Electrical energy for mixing the HRP will be generated from methane produced from fermenting algae in the methane fermentation system. These systems will consist of deep covered ponds equipped with provisions for gas capture, scrubbing, and utilization for power generation. Algae have an energy content of about 6.5 kWh/kg AFDW. During long-term mesophilic fermentation, 45% of this energy will be extracted in the form of methane, and 30% of the energy in methane can be converted to electrical energy in a heat-power generator.[14,15] The result should be that each kilogram of algae AFDW will yield 0.875 kWh of electrical energy. At an algal productivity of 200 kg/ha/d, the electrical energy yield would be 175 kWh/ha/d.

The electrical energy required to mix high-rate ponds at 0.5 ft/s for 24 h/d has been determined experimentally to be about 20 kWh/ha/d. This is about 12% of the energy available when the algal productivity is 200 kg/ha/d and 25% of the energy when productivity is 100 kg/ha/d.[13]

Final polishing of the reduction chamber effluent will be by DAF, using iron or alum salts as a coagulant. Part of the DAF effluent will be pressurized and recycled, demanding an additional amount of electrical energy estimated to be 50 kWh/ha/d.

Black sludge from the methane fermentation system will be pumped into the reduction chamber, where it will be retained indefinitely. Effluent from the HRP, containing some algae and other suspended solids, will pass through this reduction chamber with a hydraulic residence time of about 1 d. It is desirable that the reduction chamber be charged with HRP water only at night, when dissolved oxygen in the water approaches zero due to algal nocturnal respiration. It is expected that the reducing environment in the reduction chamber, together with iron salts, will render the selenate insoluble. At present, we do not know whether the reduction chamber should be provided with submerged surfaces or some form of mixing to enhance reduction, but we intend initially to attempt the reduction without them. If excessive amounts of reductant escape in the reduction chamber effluent, submerged surfaces may be added to retain the microbial population and associated reducing material. Effluent from the reduction chamber is expected to contain significant quantities of extremely fine suspended solids and residual algal pigments, and it is expected to be dark and somewhat odorous in appearance.

Pressurization of a portion of the DAF effluent will be with oxygen-rich air drawn from the HRP. It is expected that this pressurized water will quickly destroy sulfide odors in the stream entering the DAF unit upon recycling. A major determinant of overall cost will be the quantity of metallic salts and the degree of pressurization required to obtain a clear final effluent. These amounts can only be determined in a pilot installation larger than the bench-scale units currently used in our studies.

Surplus float from the DAF unit, likely containing iron salts and presumably insoluble selenium, will be dewatered on underdrained sand beds. Some of the iron/selenium-bearing material may be recycled to the reduction chamber to act as "seed" and to introduce iron to the unit. None of this material will be permitted to reenter the methane fermentation system, since it would likely contribute to toxicity.

Off-gases from the reduction chamber and the methane fermentation system are likely to contain hydrogen sulfide and hydrogen selenide. These will require control measures such as capture, stripping, and precipitation, thereby adding to the cost.

It should be apparent from this description that the cost of using algae as a reductant may be significant and will require careful evaluation in comparison with alternative reductants such as methanol or Steffens waste, which is an organic by-product of the local sugar beet processing industry.

E. DEPENDABILITY OF REDUCTANT SUPPLY

The fact that microalgae readily grow on drainage waters in HRPs and tend to harvest themselves by sedimentation, along with their tendency to undergo methane fermentation in a well-designed system, assures a continuous supply of material with reducing capability. Independence from an external supply of reductant is thus attained from the same system that provides energy. As noted previously, a major question does remain with regard to the presence of excessive amounts of precipitated solid materials such as carbonate and sulfate in the HRP sediment. Excessive sulfate will render the precipitated material inimical to its vital role in both methane fermentation and selenate reduction. Pretreatment of settled material to remove sulfate may thus be essential. The economy of such treatment will depend on the sludge concentrations attainable and the value of the residual solids. For example, calcium sulfate is a valuable additive to saline soils.

F. PRECIPITATION OF REDUCED MATERIAL

Following their fermentation, microalgae retain more than 50% of their reducing power. The biochemical oxygen demand (BOD) of microalgae is approximately 1.55 times their AFDW, so a 200 mg/l concentration of microalgae has a total potential reducing power of 310 mg/l. If 50% of this is consumed in methane fermentation, a balance of 155 mg/l remains. Thus, the environment in the reduction chamber will be extremely anoxic. The intent at this point is to control the degree of anoxia (i.e., the redox potential) and the pH in order to achieve selenate reduction and precipitation without initiating microbial reduction of soluble sulfate contained in the bulk of the water emerging from the HRP. As indicated previously, our preliminary experiments with fermented sewage-grown microalgae indicate that this can be achieved, but its chemistry and control require further exploration. More recent experiments using fermented algal sediments from our Mendota bench-scale plant HRP indicate an overt interference both with methane fermentation and with selenate reduction by some factor absent in sewage-grown algal fermentation sludges. The composition of algal coprecipitates and their interference with selenate removal is the major thrust of our current studies.

G. REMOVAL OF PRECIPITATES

Ongoing work with DAF using the proprietary Supracell unit manufactured by the Krofta Engineering Corporation (Lenox, MA) is highly encouraging.[16] In work at the University of California-Berkeley's Richmond Field Station, waste-grown microalgae have been harvested with the Krofta Supracell in as little as 3 min residence time, compared with 30 to 40 min residence times for other state-of-the-art DAF systems.[17] Application of these 2 l/s units to the 0.02 l/s effluent from our Mendota experimental algal production unit and reduction chamber effluents is not practical. However, bench-scale studies indicate that clarification of reduction chamber effluents can be achieved. Further studies of this factor must await pilot-scale operations.

H. LEGAL DISPOSAL OF SLUDGE

The nonalgal portion of the precipitates from the high-rate pond or from pretreatment of the sludges or waters will require disposal. We currently observe a decrease in TS during passage through the pond amounting to some 1000 mg/l. This material is believed to consist of $CaCO_3$ and $CaSO_4$, both of which are known to have value in altering sodium ratios in the soil solution of alkaline soils. This material, produced in the presence of supersaturated levels of dissolved oxygen, is not likely to contain significant amounts of selenium or other toxic elements. The material is subject to dewatering and drying on underdrained sand beds. It could then be stockpiled in the dry state and be made available free of charge to farmers as a soil conditioner.

Any excess sludge from the methane fermentation system, along with the DAF float from the final effluent, should contain enriched amounts of insoluble selenium and, thus, after dewatering and drying on underdrained sand beds, will constitute a hazardous material to be disposed of as a toxic or hazardous waste. It has been suggested that this material could be used on eastern portions of the San Joaquin Valley said to be deficient in soil selenium; however, such an application would no doubt require experimental studies and special permits.

Waters from underdrained sand beds will be recycled through the process. Final liquid effluents from the DAF units may require short-term impoundment for evaluation, but in the event that they cannot be discharged, they will require long-term evaporation involving huge amounts of land. This problem is shared by effluents from all processes currently under study. Their content of boron, cadmium, and strontium may preclude their use in irrigating any but natural halophilic plants. Proper disposal of treated drainage water by way of a master drain into the tidally flushed portions of San Francisco Bay is the best engineering solution, but it has been precluded by political decisions largely based on emotions rather than the rational application of facts. Gases from the methane fermentation system and reduction chamber will require treatment to remove acidic components prior to introduction to the heat-power generator. Combustion gases may also require treatment prior to their recycling for carbonation in the HRP. Such treatment of gases is a well-established technology, but, of course, will add to the cost of the overall process.

III. BENEFITS OF ALGAL SYSTEMS

Aside from their use as a reductant, algae have a number of unique potential uses either in the system depicted in Figure 1 or in other systems discussed in this book. For example, since they raise the pH of the water in which they are growing, algae soften water by precipitating Ca^{2+}, Mg^{2+}, and other polyvalent metals. Since this does not require the addition of chemicals or the use of resins, it is an economical way of removing hardness ions and heavy metals that interfere with reverse osmosis and distillation processes. Algae utilize nitrate in their biomass protein production. The proportion is about one part of nitrate-N for every ten parts of biomass AFDW.[18] Thus, any selenium reduction process could benefit from pregrowth of algae.[19] Algal biomass can be fermented to produce methane, which can be used as a fuel to generate electricity, thereby making treatment systems more or less self-sufficient in their electrical energy needs. While algal systems require land, the amounts are small compared with the enormous areas required to evaporate treated wastes, as now appears likely since no surface discharge is feasible or permissible in the San Joaquin Valley. The algal growth area occupied by an HRP could contribute significant evaporation, actually at a higher rate than deeper ponds. Due to shallow depths and to paddle-wheel mixing, the wetted areas exposed to air are greatly increased over those of deeper, unmixed ponds, so evaporation is proportionally increased. In the event that drainage water evaporation is a continued practice in the valley, some waters will reach salt levels favoring growth of *Dunaliella*, a salt-tolerant alga that is receiving much attention as a source of glycerol and pharmaceutical products.[20] *Spirulina*, a blue-green alga (actually a cyanobacterium) renowned as a future world protein source, grows naturally in the San Joaquin Valley in alkaline and saline beds. These organisms are more fermentable than the green algae (up to 70%) due to their thin cell walls.[21] Unfortunately, *Spirulina* is not as productive as the ''algae weeds'' such as the thick-walled *Scenedesmus* that grow in the HRP.

Algae growing naturally and often superficially in large, unmixed ponds cannot be managed or harvested satisfactorily. They often undergo thermal death and windrowing due to stratification, and, as a result, they create odors and attract flies. To gain their many potential benefits, algae must be grown in the specially designed HRPs. In these units, algae can be grown in fairly predictable concentrations and with fairly predictable productivities. Through the use of paddle-wheel mixing, their premature flotation or sedimentation can be largely precluded. Paddle wheels require little energy for their operation and often aid not only evaporation, but also separation by accentuating the natural aggregation of algal-bacterial flocs that would be disrupted or never formed with other methods of mixing.[13] Algal-bacterial slimes quickly form a sealant for soils lining aerobic HRPs, and as long as the soils remain submerged, the sealant remains intact. Thus, the fabricated lining of HRPs may prove unnecessary. This sealing effect due to algal-bacterial slimes is, of course, well known to those who have attempted to remove algae from water by filtration. These biological slimes also have the unique property of lowering the frictional resistance of channels. In Concord, CA HRPs, Mannings ''n'' values as low as 0.008 were estimated based on head-loss measurements in a mile-long earthwork channel.[21]

Such ancillary benefits of microalgae may seem trivial, but they could amplify into many millions of dollars in savings for districts required to treat their drainage waters prior to disposal.

ACKNOWLEDGMENTS

This research has been supported through contracts and grants by the California DWR, the San Joaquin Drainage Program of the U.S. Department of Interior, and the University of California. We are indebted to Dr. Randolph Brown and Dr. Edwin W. Lee for their continued interest and support. We also wish to acknowledge Microbio Resources of San Diego for their loan of paddle wheels and pond-lining materials; Dr. Milos Krofta of the Krofta Engineering Corporation of Lenox, MA for loan of the pilot DAF system; Enka America, Inc. of Asheville, NC for their porous tubing used for carbonation; Murietta Farms of Mendota, CA, where our algal production unit is located; and Messrs. Rodney C. Squires and Desmond W. J. Hayes and Ms. Mary Pederson of Binnie California, Inc. for their invaluable continuing assistance with the operation of our small-scale system in Mendota. We also wish to acknowledge Eleanor L. Poirier and Elizabeth D. Waiters for typing this manuscript.

REFERENCES

1. **California State Water Resources Control Board,** Regulation of Agricultural Drainage to the Sacramento River, Draft rep., CSWRCB, Sacramento, 1987.
2. **Lee, E. W.,** personal communication, 1987.
3. **Pintner, I. J. and Provasoli, L.,** Heterotrophy in Subdued Light of 3 *Chrysochromulina* Species, Haskins Laboratories, New York, 1968, 25.
4. **Lindstrom, K.,** Selenium as a growth factor for plankton algae in laboratory experiments and in some Swedish lakes, *Hydrobiologia,* 101, 35, 1983.
5. **Wrench, J. J.,** Selenium metabolism in the marine phytoplankters *Tetraselmis tetrathele* and *Dunaliella minuta, Mar. Biol.,* 49, 231, 1978.
6. **Patrick, R.,** Effects of trace metals in the aquatic ecosystem, *Am. Sci.,* 66, 185, 1978.

7. **Shrift, A.**, Biological activities of selenium compounds, *Bot. Rev.*, 550, 1958.

8. **Arthur, J. F.**, personal communication, 1985.

9. **Brown, R. L.**, The occurrence and removal of nitrogen in subsurface agricultural drainage from the San Joaquin Valley, California, *Water Res.*, 9, 529, 1975.

10. **Wang, H. K. and Wood, J. M.**, Bioaccumulation of nickel by algae, *Environ. Sci. Technol.*, 18, 106, 1984.

11. **McKeown, B. and Marinas, B. J.**, The chemistry of selenium in an aqueous environment, in *Selenium in the Environment*, Publ. No. CAT1/860201, California Agricultural Technology Institute, Fresno, July 1986.

12. **P.R.C., Inc.**, Los Banos Demonstration Desalting Facility Final Engineering Report: Physical/Chemical Pretreatment and Reverse Osmosis Desalting Systems, prepared for the State of California Department of Water Resources, P.R.C. Engineering, Inc., Orange, CA, 1983.

13. **Oswald, W. J.**, Large-scale culture systems: engineering aspects, in *Micro-algal Biotechnology*, Borowitzka, M. and Borowitzka, L., Eds., Cambridge University Press, London, 1988, 357.

14. **Golueke, C. G., Oswald, W. J., and Gotaas, H. B.**, Methane fermentation of algae, *Appl. Microbiol.*, 5, 47, 1957.

15. **Eisenberg, D. M., Oswald, W. J., Benemann, J. R., Goebel, R. P., and Tiburzi, T. T.**, Methane fermentation of microalgae, in *Anaerobic Digestion*, Stafford, D. A., Wheatley, B. I., and Hughes, D. E., Eds., Applied Science Publishers, London, 1979, 99.

16. **Nurdogan, Y. and Oswald, W. J.**, An Advanced Dissolved-Air Flotation System for Microalgae Separation, Rep. 87-4, Sanitary Engineering and Environmental Health Research Laboratory, University of California, Berkeley, 1987.

17. **Parker, D. S.**, Performance of alternative algae removal systems, in *Ponds as a Wastewater Treatment Alternative*, Gloyna, E. F., Malina, J. F., and Davis, E. M., Eds., University of Texas, Austin, 1978.

18. **Shelef, G., Oswald, W. J., and Golueke, C. C.**, Assaying algal growth with respect to nitrate concentration in a continuous flow turbidostat, in *Advances in Water Pollution Research*, Vol. 3, Jenkins, S. H., Ed., Pergamon Press, Oxford, 1971, 25.

19. **Beck, L. A. and Oswald, W. J.**, Nitrate removal from agricultural tile drainage by photosynthetic systems, in *Sanitary Engineering Research, Development, and Design*, American Society of Civil Engineers, Ithaca, NY, 1969, 41.

20. **Ben-Amotz, A. and Avron, M.**, Glycerol and beta-carotene metabolism in the halotolerant alga *Dunaliella*: a model system for biosolar energy conversion, *Trends Biochem. Sci.*, 6, 297, 1981.

21. **Oswald, W. J., Meron, A., and Zabaht, M.**, Designing waste ponds to meet water quality criteria, in *Proc. 2nd Int. Symp. Waste Treatment Lagoons*, University of Kansas, Manhattan, 1970, 186.

Chapter 10

THE ENGINEERING OF MICROALGAE MASS CULTURES FOR TREATMENT OF AGRICULTURAL WASTEWATER, WITH SPECIAL EMPHASIS ON SELENIUM REMOVAL FROM DRAINAGE WATERS

Gedaliah Shelef

TABLE OF CONTENTS

I. INTRODUCTION

During the Symposium on Biological Treatment of Agricultural Wastewater, held at the Scripps Institution of Oceanography, La Jolla, CA, from August 3 to 5, 1987, the Working Group on the Engineering of Mass Cultures of Algae held deliberations which are summarized in this chapter. The members of this Working Group were Dr. Mark Huntley, Scripps Institution of Oceanography, La Jolla, CA; Dr. Edwin W. Lee, U.S. Bureau of Reclamation, Sacramento, CA; Mr. Thomas W. Naylor, Microbio Resources, Inc., Calipatria, CA; Dr. Arthur M. Nonomura, Microbio Resources, Inc., San Diego, CA; Dr. William J. Oswald, University of California, Berkeley; and Dr. Gedaliah Shelef, Technion, Israel Institute of Technology, Haifa (Chairman).

Algal mass cultures high-rate ponds (HRPs) can be used for the treatment of various agricultural wastes, ranging from animal manures (cattle, poultry, piggeries, dairy cows, etc.) to agro-industrial wastes (pineapple processing, fruit and vegetable canning, etc.) to slaughterhouse and meat-packing wastes and other organic wastes.

The use of algal HRPs for the above wastes preferably can be preceded by anaerobic digestion to knock down the high concentrations of organic matter and to produce biogas as a source of energy.[4] The advantages of an algal treatment in this case would be using photosynthetic oxygenation to save energy in aeration, to remove nutrients, and to produce a proteinaceous by-product to be used as animal feed.

The workshop members decided that this chapter should not include the above algal treatment of agricultural wastes in order to prevent expanding the scope of this work, thus limiting it solely to the treatment of agricultural drainage waters. These waters do not contain significant concentrations of organics, but are typified by high concentrations of salinity and of nutrients, particularly nitrogen. They might contain inorganic compounds such as selenium and boron, as well as heavy metals (cadmium, strontium, molybdenum, etc.) and recalcitrant organic compounds.

In the realm of agricultural drainage waters, the engineering of mass cultures of algae to remove nitrate, as demonstrated in Firebaugh in the San Joaquin Valley of California during the late 1960s and early 1970s,[1,2] although serving the workshop as an important source of data, was also not included in the scope of the workshop deliberations and, hence, not in this chapter.

The process for the removal of selenium from subsurface tile drainage, developed at the University of California by Oswald et al.,[3] served as the sole basis for this chapter dealing with the engineering aspects of microalgal processes. This process introduced a combination of microalgal photosynthetic production of biomass together with the reducing power of anaerobic material digestion of the algal biomass, thus providing chemical reduction of soluble selenate (SeO_4^{2-}) to metallic selenium, selenite (SeO_3^{2-}), or other selenium-rich insoluble compounds. Concentrated sludge containing the precipitated insoluble selenium-rich compounds should be disposed of in a nearby hazardous-waste disposal site.

II. THE ENGINEERING DESIGN PARAMETERS AND CAPITAL COSTS

A. PRINCIPAL FEATURES OF THE PROCESS

The detailed description of the process aimed at the removal of selenium at concentrations ranging from 100 to 1400 (average 325) μg/l in raw subsurface tile agricultural drainage water to less than 10 μg/l, which is considered environmentally safe in receiving bodies of water, is given in Chapter 9 of this book.

Figure 1, drawn from the above chapter, describes the main features of the process, which are as follows:

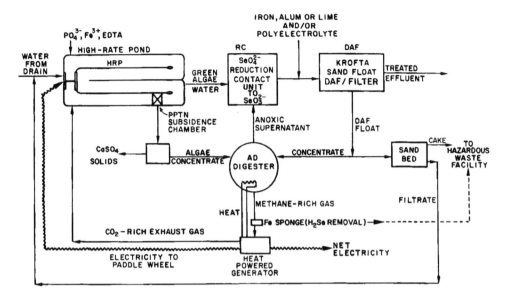

FIGURE 1. Iron, alum of lime, and/or polyelectrolyte. (From W. J. Oswald, University of California, Berkeley.)

1. The algal high-rate pond (HRP)
2. The anaerobic digester (AD)
3. The reduction contact unit (RC)
4. The dissolved air flotation (DAF) unit (combined with filter)

The principal engineering design parameters, capital investment, and operations costs, as elucidated by the workshop members, are analyzed in order to estimate the overall cost of the process and to evaluate its economical feasibility.

It should be emphasized that the engineering design parameters and the unit costs are rough estimations to achieve the general feasibility of the proposed process.

B. THE ALGAL HIGH-RATE POND (HRP)

The HRP should photosynthetically produce microalgal suspended biomass (most likely of the species *Scenedesmus*) using subsurface tile drainage water rich in nitrogen as the culture media. Phosphorus, CO_2, and, initially, iron, as well as EDTA, should be added as nutrients. The pond is mixed by paddle wheels and includes a precipitation-subsidence chamber to produce an algae-rich concentrated slurry to be fed into the anaerobic digester. Removal of nitrogen by incorporation into the algal biomass, as well as precipitation of calcium, magnesium, and some heavy metals, are some added advantages of the HRP.

Design parameters include:

- Design flow of drainage waters: 10 million gal/d (MGD)
- Retention time: 3 d (based on inflow)
- Pond depth (wet): 30 cm (approximately 12 in.)
- Expected average algae concentration: 200 mg/l
- Average algal biomass productivity: 20 g/m²/d
- Area requirements: 37.8 ha or 93.5 acres
- Design pond area: 100 acres
- Ten raceway-type ponds with earthwork dividers (car width), each pond having a 10-acre area
- Channel width: 40 ft (approximately 12.6 m)

- Longitudinal flow: 20 cm/s
- Two paddle wheels of 20-ft width and 6-ft diameter at 6 rpm for each 10-acre pond
- Heat loss in each pond (10,000-ft channel length): estimated at 6 cm (0.1 ft/mi length)

Ponds should not be lined, and only the wet side of the earthwork dividers should be covered with gunnite. Each pond should include a precipitation-subsidence chamber producing an algal slurry of between 2 and 3% solids after 1 h of thickening, assuming 70% removal of the suspended matter.

Capital costs (in 1987 U.S. dollars) include:

- Laser leveling (both rough leveling and fine grading) @ $5,000 per acre or $50,000 per pond — $500,000 for 10 ponds
- Gunnite @ $1/ft^2 and 5,000 ft^2/acre or $50,000 per pond — $500,000 for 10 ponds
- Inlets and outlets, subsidence chamber, slurry pumps and piping, electricity, and miscellaneous @ $150,000 per pond — $1,500,000 for 10 ponds
- Twenty paddle wheels (two per pond) @ $25,000 each — $500,000 for 10 ponds
- Total capital investment in HRP — $3,000,000

C. THE ANAEROBIC DIGESTION SYSTEM (AD)

This is a mesophilic digester (probably to be constructed in multiple units) with a 40-d retention time aimed at providing a chemically reduced slurry for the reduction contact unit, as well as providing biogas (methane) to supply the energy (fuel) demands of the overall scheme.

- Slurry flow: 70,000 gal/d
- Hydraulic retention time: 40 d
- Total wet volume: 2.8 million gal
- Area: 0.5 acres
- Lining and gas holding of the earthwork digester made of PVC (50 mil thick)
- Total capital cost (in 1987 U.S. dollars) of the digestion system: $200,000
- Cost (1987 U.S. dollars) of biogas electrical generator (150 kW): $150,000
- Total cost, 1987 U.S. dollars: $350,000

D. THE REDUCTION CONTACT UNIT (RC)

This unit should provide the reduction of soluble selenate (SeO_4^{2-}) to insoluble selenite (SeO_3^{2-}), metallic selenium, or other precipitous selenium-rich compounds by mixing HRP effluent rich in soluble selenate with the AD anaerobically reduced digested slurry.

- Hydraulic retention time: 1 d
- Volume: 10 million gal
- Depth: 20 ft
- Estimated cost (1987 U.S. dollars): $600,000

E. THE DISSOLVED AIR FLOTATION (DAF)

The purpose of the DAF/filter unit is to remove residual algae-borne suspended matter, residual suspended anaerobic bacteria, and residual suspended or colloidal chemically precipitated compounds (including insoluble selenium ones) from the final effluent of the process.

The DAF float is to be filtered on a sand bed, and the cake should be disposed of at a hazardous-waste facility. Part of the DAF float can be recycled into the AD unit, while the filtrate will be recycled into the HRP.

A proprietary DAF unit manufactured by the Krofta Engineering Corporation of Lenox, MA was found by Oswald to be most efficient and cost saving.

- Hydraulic retention time: 5 min
- Total estimated cost of DAF for a 10-MGD flow: $1,000,000

F. TOTAL CAPITAL INVESTMENT

The total estimated capital investment (in 1987 U.S. dollars) for the system designed to remove selenium from subsurface tile drainage can be determined as follows:

- HRP $3,000,000
- AD and generator   350,000
- RC   600,000
- DAF   1,000,000
- Holding ponds, sand bed, and miscellaneous   150,000

- Total capital investment $5,100,000

III. OPERATION AND MAINTENANCE

Operational and maintenance costs (in 1987 U.S. dollars) for the selenium removal system are estimated as follows:

- Capital investment annualized for 20 years at 4% interest $510,000/year
- Operating labor (seven workers @ $50,000)   350,000/year
- Energy and CO_2 (to be supplied by biogas as fuel)   —
- Phosphate, iron, and EDTA   100,000/year
- Flocculants (FeCl @ 60 mg/l @ $150/ton)   150,000/year

- Total operation and maintenance $1,110,000/year

IV. CONCLUSIONS

An annual operating cost (including investment return) of $1,110,000 (U.S., 1987) to treat 10 MGD of subsurface tile drainage waters for the removal of selenium to below 10 µg/l constitutes a cost of about $100/acre-ft of drainage water. This is within the "ball park" of the costs which are feasible to farmers (probably with some governmental subsidies), ranging, according to the U.S. Bureau of Reclamation,[5] between $50 and $100/acre-ft (U.S., 1987). Economy of scale should further reduce the costs when larger schemes are built to remove selenium from all the drainage waters in the affected area in the San Joaquin Valley.

The preliminary feasibility analysis encourages bringing the microalgal/anaerobic bacterial process to a demonstration-scale facility (possibly with a 0.5-MGD flow) which will provide more accurate engineering and cost parameters.

REFERENCES

1. **Brown, R. L.,** The occurrence and removal of nitrogen in subsurface agricultural drainage from San Joaquin Valley, California, *Water Res.,* 9, 529, 1975.
2. **Goldman, J. C., Beck, L. A., and Oswald, W. J.,** Nitrate removal from agricultural tile drainage by photosynthetic systems, in Proc. 2nd Annu. Symp. Sanitary Engineering Research, Development and Design, Cornell University, Ithaca, NY, 1969.
3. **Oswald, J. W., Chen, P. H., Green, F. B., Gerhardt, M. G., Nurdogan, Y., Newman, R. D., and Shown, L.,** The role of microalgae in removal of selenate from subsurface tile drainage, in Proc. Symp. Biological Treatment of Agricultural Wastewater, Scripps Institute of Oceanography, La Jolla, CA, August 3 to 5, 1987.
4. **Shelef, G., Kimchie, S., and Tarre, S.,** Anaerobic digestion of agricultural wastes in Israel, in *World Energy Conference (WEC) Report on Biogas in Rural Areas,* Zhao, Y., Ed., WEC Publications, London, 1987.
5. **Lee, E. W.,** Introduction to the problem of selenium in the California San Joaquin Valley drainage waters, in Proc. Symp. Biological Treatment of Agricultural Wastewater, Scripps Institute of Oceanography, La Jolla, CA, August 3 to 5, 1987.

Chapter 11

IS THERE AN "UNCERTAINTY PRINCIPLE" IN MICROBIAL WASTE TREATMENT?

James W. Blackburn

TABLE OF CONTENTS

ABSTRACT

Predictions of the performance and fate of chemicals in mixed-culture microbial waste-water treatment systems have been notoriously inexact. Studies of the treatment and fate of difficult-to-degrade chemicals at different times or on different scales often do not give similar results. Basic engineering questions related to the time required for treatment and the final effluent concentration of full-scale processes cannot now be reliably answered for the biological treatment of many environmentally significant chemical compounds, even when numerous prior laboratory or pilot studies have been implemented. The design and performance of full-scale systems may not be reliable, and more costly treatment options may be chosen in light of these uncertainties. This chapter explores the sources of these uncertainties and the potential for increased reliability based on emerging experimental technology.

I. INTRODUCTION

Agriculturally related process and wastewaters are becoming more significant as sources of surface and groundwater pollution. In addition to a potentially high inorganic nutrient content, these waters may contain a diverse and widely varying set of organic agricultural chemicals. Treatment leading to disposal or reuse of such waters often requires the removal of these compounds, and biological treatment is often considered. The nature of the chemicals, the spatial and temporal variations in concentration and composition, and the need for low effluent concentrations make this application a challenging one. With an eye to the treatment of agricultural wastewaters, this chapter will address the general problem of mixed microbial treatment of liquid and solid wastes containing organic compounds that only biodegrade slowly or are otherwise difficult to degrade.

Researchers and practitioners involved in the application of mixed-culture microbial systems to waste treatment and bioprocessing have long been aware of the complexity of such systems.[1-4] Research has often followed the lines of classical microbiology, with emphasis on pure cultures of isolates from mixed environmental populations.[2] Development and application efforts have often focused on unstructured models of both the organic substrate and the biomass, leading to the testing and development of technology emphasizing distributed organic removal by a biomass assumed to be well represented by distributed measures.[5,6] This work has encouraged the understanding of biodegradation to grow, while allowing the scaleup of biotreatment processes capable of reliable overall organic removal (often measured as biochemical oxygen demand [BOD], chemical oxygen demand [COD], or total organic carbon). Process models and kinetics have been expressed in terms of these overall parameters, and "treatability" studies have been routinely performed on each waste stream to support system design.

Pure-culture studies, often with sole carbon sources, have resulted in the biochemical characterization of degradation for many compounds. Thousands of papers have been published in this area, and reviews are available.[7-11] This information has sometimes been coupled with genetic studies, leading to the identification of the genes responsible for the catabolism of given compounds as well as the genetic regulation and control of biodegradation at the molecular level. The wealth of knowledge building in both the theory and practice of mixed-culture bioprocessing is extensive, but unfortunately not very useful in achieving a given level of treatment for specific compounds in a given waste stream or bioprocess feed. Recent experiences have shown that the ability to predict system behavior and performance for a given design and set of operation conditions is very limited, even with numerous prior laboratory studies. Examples of this limitation include cases where

1. Compound fates established in laboratory studies may be substantially different than in pilot-scale, full-scale, or even other laboratory studies.[12,13]

2. Positive biotransformation results from small lab-scale systems are often not reproduced in different systems or even in the same systems at different times.[14-17]

3. When a given biotransformation is achieved at a larger scale, the kinetics are often unrelated to distributed measures of biomass.[18]

4. Instantaneous biotransformation rates vary widely and in an apparently stochastic manner, even in "well-operated", "steady-state" systems. Variability of biotransformation rate constants for specific compounds is often large, sometimes exceeding three orders of magnitude.[18-21]

5. Effects of competing fate mechanisms such as stripping/volatization, sorption, and chemical reactions are not well understood at any scale, and reported biological removal of a compound often includes removal/conversion related to competing abiotic processes. Comparisons of kinetics across systems and waste streams are confounded.[12,18]

6. Competing mechanisms may be quite configuration specific and likely will scale much differently than the biotransformation mechanism.[12,19,22,23]

Today, especially in waste treatment, it is clear that although laboratory testing and design for overall carbon removal is a science, a similar effort with the goal of biotransformation of a specific organic compound is an art. Success in implementing a full-scale project for biodegradation of a specific, difficult-to-degrade compound with designs based on lab- or pilot-scale results is fortuitous.

Practitioners are generally unable to consistently predict either the final concentration or the time of treatment for specific, difficult-to-degrade compounds. The present practice is to extensively overdesign systems and avoid making performance warranties for specific compound removals. All of these results contribute to a view of mixed-culture biological treatment as highly stochastic or chaotic, if not downright mystical, even though all microbiological, biochemical, and genetic sciences point unflinchingly to highly deterministic and highly regulated bioprocesses.

If the scaleup of pure- and/or mixed-culture studies using unstructured models is uncertain and if reliability for waste processing is important, then a modified approach must be developed to provide reliable predictions and designs.

It is tempting to consider classical engineering approaches for deterministic process description. Here, a process model is proposed and simplified to a level where parameters can be defined and experimentally determined. The model is then validated for a set of conditions and is used to project performance for different conditions. Unfortunately, results from this approach have rarely been fruitful for predicting difficult-to-degrade compound fates in biological treatment.

Many other examples in engineering also exist where predictive determinism fails. Even in that "purest" of experimental disciplines, physics, the famous Heisenberg Uncertainty Principle reminds us that there are always limits to our ability to deterministically describe and measure physical systems. To paraphrase this axiom, the act of measuring a particle's mass disturbs the system and increases the uncertainty of knowing the particle's momentum.

Determinism in mixed-culture biological treatment is confounded both by the level of complexity and by a metaphor for Heisenberg's principle. With conventional technology, our attempts to infer the state of mixed-culture biotransformations using measurements from samples taken from the system depend upon the ability to preserve all sample biochemical and ecological characteristics and the ability to measure those characteristics without bias from the chemical, biochemical, or microbiological analytical techniques. In the general case, off-line sampling and analysis may alter both the community structure and the resulting degradation kinetics of interest and increase the predictive uncertainty in both the form of the model and the state of important kinetic parameters of the operating system. In other words, the actual state of the operating system may not even resemble that predicted from the state of the sample. Presently we cannot directly measure the operating system, and the precision of the sample's simulation of the system cannot even be resolved.

This limitation is felt most seriously in the discipline of microbial ecology and has been discussed extensively.[24] For example, we have documented the loss of phenotypical activity in the biotransformation of phenol in activated sludge samples taken from an operating system; the samples were rapidly cooled on ice and challenged with phenol within several hours using a highly controlled batch assay.[12,20]

II. A SYSTEMS VIEW OF THE UNCERTAINTIES

To further examine the uncertainties of mixed-culture systems, it is useful to realize that a given mixed-culture system may be described at a minimum of four distinct levels (Figure 1):

1. Reactor level: the macro view of the system that recognizes overall system characteristics (configuration, residence times, etc.), advection of components, competitive fate processes, and other macro-scale issues
2. Ecological level: the level that deals with groups, guilds, or subpopulations of microorganisms, their interactions with their environment (effects of mixing, power input, physical habitats, etc. on the groups, guilds, etc.), and their self- and cross-interactions
3. Cellular physiological level: the level that deals with the behavior of cells based on taxa or phylogeny, growth and biotransformation rates of each cell class, and events that affect the general growth or death of all microorganisms, etc.
4. Molecular level: the level at which the molecular processes of gene replication, maintenance, transcription, translation, protein synthesis, enzyme activation and inhibition, and other molecular processes can be examined

A. REACTOR LEVEL

The reactor level is the level most often studied and analyzed by engineers. At this level, systems are designed, operated, and controlled, and often a mathematical model of suitable complexity is used to aid in system analysis. Knowledge of transport processes, flow, mixing intensity, alternative fate processes (stripping, sorption, etc.), presence of terminal electron acceptors (O_2, nitrates, sulfates, etc.), and other variables is incorporated if and only if reliable information regarding the form of the model and related empirical parameters is available. Necessary parameters for development of reliable predictive performance models may include physical/chemical properties of the fluids and solids, chemical analysis of the compounds in the fluid and solid phases, microbiological analysis of the fluid and solid phases, and other parameters. Problems exist in the sampling and measurement of many of the critical parameters needed for a reliable predictive model.

Heterogeneity in mixed-culture systems implies spatial variations and leads to increased sampling complexity. This is particularly troublesome in soil systems, where uneven distribution of hydraulic, chemical, physical, and biological variables is certain and where impractical numbers of wells, samples, and analyses are therefore required.

Proof of biotransformation (vs. compound disappearance by other fate mechanisms) requires chemical identification of a metabolite or a mass balance with a convincing estimation of alternative fate pathways. It can be shown from biochemical concepts and in numerous controlled studies that such intermediate metabolites appear at vanishingly low concentrations, and even if they are measurable, the uncertainty of the analysis is large. Use of radiolabeled tracers is generally acceptable only in lab or controlled pilot studies, and proof of biotransformation is more elusive at larger scales.[12]

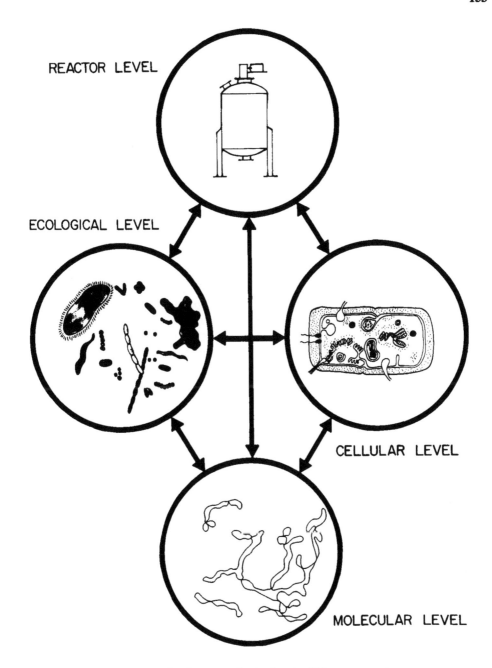

FIGURE 1.   The four system levels of a microbial treatment system.

System transients may be important in the induction of system instability,[25] but difficult to detect and model. The shortest detectable transient that can be seen by sampling the aqueous phase in a microbial system is on the same order as the reactor hydraulic residence time, and dynamics of faster transients and resulting effects may be lost. Fast, undetectable transients that may occur may still have the potential for major system upsets. Off-gas samples may provide insight into shorter transients influencing volatile components,[21] but will probably not enlighten us regarding the dynamic behavior of nonvolatile metabolites.

The form of the kinetic model for a specific biotransformation cannot even be determined *a priori*. Many types of biotransformation models exist, but are normally selected by fitting experimental data after the fact.[3] Where lab-scale studies are used to predict full-scale systems, it is assumed, but not always proven, that lab-scale kinetic models apply to the full-scale system.

Perhaps the area of greatest uncertainty is the community structure of the microorganisms in the system. Because of difficulties inherent in the sampling and analytical procedures for characterization of various subpopulations in a microbial biomass,[24] measurement of distributed parameters such as total suspended solids and light transmittance has been the preferred approach. Distributed biomass parameters disclose little information about the microorganism subpopulations responsible for specific compound degradation in mixed cultures. Models based on distributed measures may be highly inaccurate.[19-25]

B. ECOLOGICAL LEVEL

The ecological level has rarely been quantitatively considered because, until recently, the ability to structure the biomass into degradative subpopulations and formulate suitable rate expressions has been satisfactory only for the most elementary of microbial systems. Analogous to Heisenberg's principle, available sampling and analytical methods have often led to modification of system structure and kinetics. With no basis for measuring the size of degradative subpopulations, any approach to assigning activity to the "correct" subpopulation and developing reliable kinetic expressions is doomed.

Subpopulations of organisms interact at the ecological level, and efforts to elucidate the interactions between subpopulations and their environment have been confined to contrived, gnotobiotic systems. Furthermore, ecological interactions in mixed-culture microbial systems are structurally complex. To illustrate the experimental difficulties related to complexity at the ecological level, consider a hypothetical approach to developing data for a system's community structure and interactions. A mixed-culture system containing numerous ecological interactions has been fully characterized to determine the number of community members, and each has been isolated. Growth characteristics of each community member are then studied, both in monoculture and in binary culture with all other members and at constant conditions simulating the treatment process. The effects of self- and cross-interactions on the growth of each member are quantified, and a deterministic ecological model is constructed incorporating the interaction information. The populations of a member or members responsible for the biotransformation of a specific compound are predicted and incorporated into a kinetic expression for estimating treatment performance.

In this example, n^2 monoculture and binary experiments are run at each set of reactor conditions of interest. However, the ecological interaction parameters determined in a given test may also be influenced by the presence of a third, fourth, fifth, or any combination of other members, although these experiments have not been performed.

If one could measure all self- and cross-interactions (interactions in the presence of all combinations of all community members), the number of self-interactions in a community would be

$$\text{Number of self-interactions} = 2^{n-1} \cdot n \tag{1}$$

The number of cross-interactions would be

$$\text{Number of cross-interactions} = 2^{n-1} \cdot \left(\frac{n!}{2(n-2)!} \right) \tag{2}$$

Note that the term 2^{n-1} arises because member extinctions are permitted and numerically different interactions between the remaining members can occur.

TABLE 1
The Ecological Confoundation Factor

n	p	C
1	1.0	1:1
2	0.67	1.5:1
3	0.38	2.7:1
4	0.20	5:1
5	0.10	9.6:1
6	0.054	19:1
7	0.027	37:1
8	0.014	72:1
9	0.007	142:1
10	0.003	282:1

Note: n = number of community members; p = probability of reliable prediction with known information; C = ecological confoundation factor — the odds against reliable prediction of community behavior, even when all community members and all member self- and binary interactions are known.

The probability, p, that predictions using self- and binary interactions are correct in the general case would be

$$p = \frac{2^{n-1} \cdot \left(n + \dfrac{n!}{2(n-2)!}\right) - n^2}{2^{n-1} \cdot \left(n + \dfrac{n!}{2(n-2)!}\right)} \tag{3}$$

The reciprocal of p is C and is shown numerically in Table 1. C represents the odds against a correct prediction of community behavior when all community members and all self- and binary interactions are experimentally known. C is an ecological confoundation factor. Even when knowledge about the community and system is extensive (n^2 experiments), little can be reliably said of the interactions of the community itself because interactions in the test systems may be numerically different from those in the environmental system. From a practical standpoint, unless only a few interactions are ever important and these are known *a priori*, the ecological level in a working mixed-culture system is indeterminant.

C. CELLULAR LEVEL

At the cellular level, processes that regulate and control overall cell growth are numerous.[25] When a cell is under balanced growth conditions, thousands of metabolic processes may proceed and the cell must respond to environmental changes using internal biochemical control mechanisms. In addition, the cell has the capacity to store critical material for times when it would otherwise be growth limited. There may be, depending on the cell's recent past history, a time lag between the time the cell can no longer manufacture a critical material and the time when cellular growth is affected. During this interval, no noticeable change in cellular growth may be evident. Furthermore, the original cause of unbalanced growth may be obscured to the researcher because of the time lag. Finally, unbalanced growth can occur even if an inventory of critical material exists, especially if biotransformations are required to convert the stored material to a usable form. In this case, a transient period of unbalanced growth may occur until the appropriate biochemical pathways are induced.

Unbalanced growth in a given type of cell may negatively affect degradation of an organic compound. However, other cellular strategies may exist where the organism does not need to employ catabolism of the specific compound while it adjusts to the new balanced growth state.

Many possibilities then exist as to the cellular response to a change in environmental or ecological conditions. These include

1. Maintenance of unbalanced growth
2. Reversible unbalanced growth — either steady or transient
3. Irreversible unbalanced growth — either steady or transient
4. Cell maintenance, but no growth
5. Reversible cell dysfunction and/or nonviability — either steady or transient
6. Irreversible cell dysfunction and/or nonviability — either steady or transient
7. Cell death

The resulting catabolic (biodegradative) or anabolic cellular process of interest may proceed (1) steadily maintained, (2) steadily disengaged, or (3) unsteadily maintained.

Nearly all combinations of overall physiological responses and specific metabolic activities are possible, with the exception of overall cell death and maintenance of activity. Even here, when one considers autolytic processes (breakdown of dead cell mass), metabolism in a sense continues even after cell death.

Knowledge of the cell's physiological processes comes with difficulty. A large fraction of all research efforts in biochemistry and microbiology are dedicated to elucidating these processes and their regulation. Emphasis has been placed on understanding mechanism forms rather than on process dynamics. Even if the dynamics were known for a given organism in axenic culture, extracellular pools of materials, dynamics of disturbances, and ecological factors all strongly influence the dynamic cellular state. In other words, traditional studies may lead to an estimate of optimal balanced growth, but little information is gained on the sensitivity and stability of the organisms to transient reactor- and ecological-level conditions in real engineered and natural systems.

Even with the extensive literature available, critical information on physiological process dynamics is lacking. Classical testing with batch or "steady-state" experiments alone cannot provide the needed dynamic response information.

D. MOLECULAR LEVEL

In recent years, great advances have been made through the study of important physiological processes at the molecular level. Instructions for cellular construction of biochemicals are stored in cellular DNA in sequences of nucleotides called genes. A cell or population of cells possessing a given gene or genes is called a genotype. Cellular regulation processes dynamically select the genes to be activated, and messenger RNA (mRNA) transcripts are synthesized. Simultaneously, enzymes that degrade the mRNA (RNAses) are present, and the relative rates of synthesis and degradation of the mRNA determine the cellular concentration of transcripts and help define the subsequent rate of protein synthesis (for a given gene). Protein synthesis occurs when a transcript binds with a ribosome and sufficient amino acids and other constituents are available in the intracellular pool for construction of the protein. Relative concentrations of the transcripts for a given protein and all other transcripts determine the instantaneous fraction of ribosomes used to construct a given protein. Generalized protein degradation by cellular proteases is also continuous, and the intracellular concentration of the protein is mitigated by the relative rate of synthesis and the relative rates of degradation to peptides and amino acids at a given time.

Protein chemistry itself is complex, and numerous conditions may be needed in order for the protein to be active. If the protein possesses catalytic activity, the protein is an enzyme. Proteins may require linkage with other proteins or peptides in order to achieve activity. Conversely, linkage with other proteins can lead to blockage or inhibition of activity. Proteins or enzymes may require binding with other chemicals to cause either an increase or a decrease in activity. Proteins may require a certain three-dimensional conformation in order to become active. This structure can be influenced by cellular conditions like temperature and composition. Numerous mechanisms, therefore, exist that may effectively regulate protein synthesis and/or enzyme activity at the molecular level.

The cell's overall physiological state and overall instantaneous optimization strategy may lead to sufficient concentrations of active protein(s) or enzyme(s). The successful expression of the activity of a given genotype is called the cell's phenotype for that activity. Possessing a genotype for an activity is a necessary but insufficient condition for possession of the phenotype for that activity. Necessary and sufficient conditions for expression of a phenotype may include any or all of the factors just discussed. The set of necessary and sufficient conditions for phenotype expression at a molecular level is likely to be different for each genotype.

The successful expression of the genotype to an active protein or enzyme may be related to cell physiological states, intracellular pools, regulation at the molecular level, and interactions with other cellular nucleic acids and peptides. Knowledge of these biochemical processes lies in the domains of molecular biology, biochemistry, and genetics, where axenic cultures, defective mutants, and recombinants are necessary for the study. The complexity at this level can easily be as great as or greater than the complexity at the ecological level, even considering only a single genotype in a single species of organism.

A cell inherits a genotype when microorganisms divide. If the gene is chromosomal, the genotype is passed to the new cell along with the other chromosomal genome. Prokaryotic cells can also inherit a genotype from nonchromosomal or extrachromosomal DNA that has the ability to replicate. These small, circular pieces of DNA are called plasmids. They may exist in bacterial cells at low or high concentrations (i.e., from low copy numbers of just a few to hundreds of copies per cell). A widely held belief is that essential genotypes for physiological processes required for the survival of a cell under unstressed conditions are often located on the chromosome, while specific genotypes for survival under special or extreme conditions are often located on a plasmid. Molecular processes called transpositions lead to the movement of genes from plasmids to the chromosome and vice versa.

Chromosomes must be conserved among daughter cells, but the daughter cells may be able to survive without inheritance of plasmids. It has also been shown that cells containing plasmids are sometimes outcompeted by identical cells free of plasmids under balanced growth conditions, presumably because more energy is required for replication of each plasmid-bearing cell than for growth of each plasmid-free cell.

In general, then, we would expect to find highly specialized genotypes such as those for xenobiotic chemical catabolism on plasmids and, further, would expect the concentration of a genotype in an unstressed system (one not experiencing a selective pressure) to be low or to diminish with time. The dynamics of these plasmid loss processes may be long relative to the metabolic process itself.

III. OVERALL SYSTEM BEHAVIOR

Returning, then, to consideration of why mixed-culture system behavior is unpredictable across configurations, scales, and applications, biotic and abiotic processes are working, may be regulated at all levels, and may have major and strong interactions across levels. Regulation at each level operates with dynamics at different characteristic times. Figure 2 shows the extreme variability in the dynamics at all levels of mixed-culture processes.

Molecular-level regulation can operate on the order of cellular diffusion times and enzyme

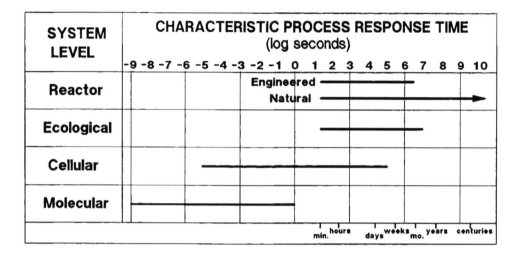

FIGURE 2.   Typical process response times at various system levels in a natural or engineered microbial treatment system.

turnover times (1 to 10^{-9} s). Cellular-level process regulation overlaps this range, but extends up to cell replication times (20 min to over 1 h) and access times for stored materials or other long-term adaptations (e.g., derepression of a biochemical pathway, selection and optimization of transport systems, or extracellular enzyme synthesis and transport). Longer-term cellular processes can be regulated at times measured in hours or days. Ecological effects can extend from minutes to days, weeks, or months. Reactor-level processes are strong functions of the reactor configuration and of "design" parameters such as mixing intensity. Engineered reactor-level processes may occur over characteristic times of minutes (for diffusion in turbulently mixed systems) to days or months. Natural reactor-level process regulation may occur from days to (in the case of low-conductivity aquifers) hundreds or thousands of years.

A simplified rule of system dynamics is that, in an elementary system, disturbances at a given time scale will induce the response of regulatory processes beginning at a similar time scale. Thus, very rapid (e.g., millisecond) transients in an important mechanism involving biodegradation of a specific compound would induce primary changes in regulatory processes operating at the millisecond time scale. In microbial systems, it is likely that some short-term changes might "cascade" through the system and cause meaningful, potentially irreversible changes in, for example, the ability of a degradative genotype to ecologically compete with other subpopulations and the subsequent local extinction of that genotype.

Most perplexing is the reality that studies done in the classical deterministic manner of controlling specific variables to emphasize "basic" relationships at all levels contribute little to our ability to make predictions of the dynamic response of the complex system. Our knowledge of the processes and mechanisms at all levels is based on "controlled studies", and since at least the biotic systems employ adaptive regulation and control, the processes can only be studied in the context of the system operating as a whole. Most of the time we have neither the knowledge nor the tools to decode or even measure the activity of interest in real systems.

We are then faced with the circular prospect that we need to study a given activity in the context of its operating complexity, but that our methods are insufficiently effective to enable system monitoring and analysis in the complex system. Practitioners in this area may attempt to study the process at the laboratory scale, the pilot scale, or in a microcosm, but often will admit that in the last analysis, large-scale application turns on the most forgiving notion of "try it and see". While it is true that successive "try it and see" testing increases the experience of a given practitioner, hope for improving design or scaleup reliability arising from these tests is an illusion when contrasted with the potential complexities.

IV. EXPERIMENTAL APPROACHES FOR REDUCING THE UNCERTAINTY

The problem of dynamic mixed-culture process uncertainties is significant and multi-faceted. Its solution may lead to major advances in the development of treatment systems and, perhaps, eventually to mixed-culture counterparts of the monoculture biochemical processes of today. Although it is impractical here to attempt to outline an exhaustive specific research agenda at all levels, some generalization can be made.

A. MOLECULAR TOOLS FOR COMMUNITY STRUCTURE ANALYSIS

Perhaps the central uncertainty problem is the inability to sample and characterize a dynamic microbial system for the microbial structure at an instant in time. The presence and number of cells related to an important activity (e.g., biotransformation of a given xenobiotic compound) vary dynamically by orders of magnitude.[12,19] Knowledge of the structure of these populations at short time intervals, coupled with measures of phenotypical activity, may provide the basis for more general kinetic relationships and greater scaleup reliability.

Attempts to develop new molecular tools for this application have recently been reviewed.[26] One promising approach is the direct quantification of the genotype and/or phenotype using labeled probes composed of the actual genes of interest.[27-29] Data on successful application of gene probes to activated sludge have been presented.[21]

Molecular tools for structuring the microbial community ideally should be sensitive, rapid, and easy to use. Widespread activity is underway to develop delivery protocols and devices for molecular tools for the purpose of quantification of a specific community member of pathogenic or medical importance, and it is likely that some of these results will be applicable to microbial waste treatment. Furthermore, work is underway to develop a device for the easy and rapid use of gene probes for the monitoring and control of waste treatment systems.[30]

B. IMPROVED LAB-SCALE EXPERIMENTAL PROTOCOLS

Since the dynamics of microbial systems are important and, from a practical point of view, indeterminant, systems analysis approaches are required to discover regimes of process operating stability and instability and to help identify the critical causative processes and mechanisms. These approaches have been applied extensively in engineering science and also on adaptive biological systems in ecology, bioprocess engineering, and biological treatment.[31-35] Problems requiring emphasis include the experimental reactor design for system data collection, the potential adaptability of microbial systems, the potential nonlinearity of the system's processes, and the extensive instrumentation required for continuous data collection of the numerous important variables.

Different reactor designs applied in parallel may lead to an improved ability to identify limiting processes and system states under steady or dynamic conditions. Adaptability of biotic process mechanisms confounds the ability to directly apply systems analysis approaches, since the perturbation required for systems analysis also could cause reversible or irreversible adaptations. Conversely, variable perturbation protocols could also allow characterization and assessment of adaptation in microbial systems. Classical systems analysis requires the assumption that the equations describing temporal variations are linear; nonlinearities preclude simple mathematical interpretations. However, the experimental perturbation response data will in itself allow definition of stability regimes and conjecture of potential causative processes. regardless of whether the data are linear. This could lead to separate experiments specifically designed to study sources of instability. The data needs for systems analysis are extensive, owing to the requirement that a large set of parameters must be sampled at frequent intervals. Specially designed facilities using sophisticated sampling and analytical instrumentation must be coupled with data reduction and analysis equipment to eludicate process relationships. These requirements are very different and much more extensive than present experimental demands, but it is anticipated that the results will be proportionally rewarding.

V. CONCLUSIONS

With present methods and experimental protocols, uncertainties in the structure and kinetics of processes in mixed-culture microbial treatment exist. The uncertainties, like the Heisenberg principle in physics, arise from the need to study a complex, undisturbed operating system and from the inability (with present methods) to accomplish this without disturbing the system. The complexity of the process at the reactor, ecological, cellular, and molecular levels and interactions across these levels make the system — for all practical purposes — indeterminant.

Unlike the Heisenberg principle, uncertainty in microbial treatment systems can be reduced with the development and application of new, emerging molecular tools and the application and refinement of systems analysis methods in conjunction with the improved tools. Although considerably greater effort will be required to successfully analyze these systems than is presently being expended in the development of biological waste treatment processes, improved understanding will lead to improved predictability, better control schemes and operability, and, possibly, a new generation of effective biochemical treatment and production processes based on controlled mixed-culture microbial systems.

REFERENCES

1. **Atlas, R. M. and Bartha, R.,** *Microbial Ecology*, Benjamin Cummings, Menlo Park, CA, 1987, chap. 4 to 6.
2. **Slater, J. H.,** Mixed cultures and microbial communities, in *Mixed Culture Fermentations*, Bushell, M. E. and Slater, J. H., Eds., Academic Press, London, 1981, chap. 1.
3. **Bazin, M. J.,** Mixed culture kinetics, in *Mixed Culture Fermentations*, Bushell, M. E. and Slater, J. H., Eds., Academic Press, London, 1981, chap. 2.
4. **Hobson, P. N.,** Microbial pathways and interactions in the anaerobic treatment process, in *Mixed Culture Fermentations*, Bushell, M. E. and Slater, J. H., Eds., Academic Press, London, 1981, chap. 3.
5. **Bader, F. B.,** Kinetics of double-substrate limited growth, in *Microbial Population Dynamics*, Bazin, M. J., Ed., CRC Press, Boca Raton, FL, 1982, chap. 1.
6. **Benefield, L. D. and Randall, C. W.,** *Biological Process Design for Wastewater Treatment*, Prentice-Hall, Englewood Cliffs, NJ, 1980, chap. 2 and 4.
7. **Gibson, D. T., Ed.,** *Microbial Degradation of Organic Compounds*, Marcel Dekker, New York, 1985.
8. **Rochkind-Dubinsky, M. L., Sayler, G. S., and Blackburn, J. W.,** *Microbiological Decomposition of Chlorinated Aromatic Compounds*, Marcel Dekker, New York, 1987.
9. **Leisinger, T., Ed.,** Reviews — microbial degradation of environmental pollutants, *Experientia*, 39(11), 1181, 1983.
10. **Leisinger, T., Cook, A. M., Hütter R., and Nüesch, J.,** *Microbial Degradation of Xenobiotics and Recalcitrant Compounds*, Academic Press, London, 1981.
11. **Chakrabarty, A. M., Ed.,** *Biodegradation and Detoxification of Environmental Pollutants*, CRC Press, Boca Raton, FL, 1982.

12. **Blackburn, J. W., Troxler, W. L., Truong, K. N., Zink, R. P., Meckstroth, S. C., Florance, J. R., Groen, A., Sayler, G. S., Beck, R. W., Minear, R. A., Yagi, O., and Breen, A.**, Organic Chemical Fate Prediction in Activated Sludge Treatment Processes, Rep. EPA 600/S2-85/102, National Technical Information Service, Springfield, VA, 1985.

13. **Raper, W. G. C.**, Important issues in the scale up of new processes, in *Scale-Up of Water and Wastewater Treatment Processes*, Schmidtke, N. W. and Smith, D. W., Eds., Butterworths, Boston, 1983, 55.

14. **Goldstein, R. M., Mallory, L. M., and Alexander, M.**, Reasons for possible failure of inoculation to enhance biodegradation, *Appl. Environ. Microbiol.*, 50(4), 977, 1985.

15. **MacRae, I. C. and Alexander, M.**, Microbial degradation of selected herbicides in soil., *J. Agric. Food Chem.*, 13, 72, 1965.

16. **Anderson, J. P. E., Lichtenstein, E. P., and Wittingham, W. F.**, Effect of *Mucor alternans* on the persistance of DDT and dieldrin in culture and in soil, *J. Econ. Entomol.*, 63, 1595, 1970.

17. **Lehtomäki, M. and Niemelä, S.**, Improving microbial degradation of oil in soil, *Ambio*, 4, 126, 1975.

18. **Feiler, H.**, Fate of Priority Pollutants in Publicly Owned Treatment Works, Rep. EPA 440/1-80/301, Effluent Guidelines Division, U.S. Environmental Protection Agency, Washington, D.C., 1980.

19. **Blackburn, J. W.**, Prediction of organic chemical fates in biological treatment systems, *Environ. Prog.*, 6(4), 217, 1987.

20. **Sayler, G. S., Breen, A., Blackburn, J. W., and Yagi, O.**, Predictive assessment of priority pollutant bio-oxidation kinetics in activated sludge, *Environ. Prog.*, 3(3), 153, 1984.

21. **Blackburn, J. W., Jain, R. K., and Sayler, G. S.**, The molecular microbial ecology of a naphthalene-degrading genotype in activated sludge, *Environ. Sci. Technol.*, 21(9), 884, 1987.

22. **Truong, K. N. and Blackburn, J. W.**, The stripping of organic chemicals in a biological treatment process, *Environ. Prog.*, 3(3), 143, 1984.

23. **Roberts, P. V., Munz, C., and Dändliker, P.**, Modeling volatile organic solute removal by surface and bubble aeration, *J. Water Pollut. Control Fed.*, 56(2), 157, 1984.

24. **Atlas, R. M. and Bartha, R.**, *Microbial Ecology*, Benjamin Cummings, Menlo Park, CA, 1987, chap. 7.

25. **Barford, J. P., Pamment, N. B., and Hall, R. J.**, Lag phases and transients, in *Microbial Population Dynamics*, Bazin, M. J., Ed., CRC Press, Boca Raton, FL, 1982, chap. 3.

26. **Jain, R. K. and Sayler, G. S.**, Problems and potential for in-situ treatment of environmental pollutants by engineered microorganisms, *Microbiol. Sci.*, 4(2), 59, 1987.

27. **Sayler, G. S., Shields, M. S., Tedford, E. T., Breen, A., Hooper, S. W., Sirotkin, K. M., and Davis, J. W.**, Application of DNA-DNA colony hybridization to the detection of catabolic genotypes in environmental samples, *Appl. Environ. Microbiol.*, 49, 1295, 1985.

28. **Jain, R. K., Sayler, G. S., Wilson, J. T., Houston, L., and Pacia, D.**, Maintenance and stability of introduced genotypes in groundwater aquifer material, *Appl. Environ. Microbiol.*, 53(5), 996, 1987.

29. **Pettigrew, C. and Sayler, G. S.**, Application of DNA colony hybridization to the rapid isolation of 4-chlorobiphenyl catabolic phenotypes, *J. Microbiol. Methods*, 5, 205, 1986.

30. **Moore, R., Blackburn, J. W., Bienkowski, P. R., and Sayler, G. S.**, Bioreactor sensors based on nucleic acid hybridization reactions, *Proc. 9th Symp. Biotechnology for Fuels and Chemicals*, Scott, C. D., Ed., Humana Press, Clifton, NJ, 1988, 325.

31. **Bender, E. A., Case, T. J., and Gilpin, M. E.**, Perturbation experiments in community ecology: theory and practice, *Ecology*, 65(1), 1, 1984.

32. **Chu, I.-M. and Papoutsakis, E. T.**, Growth dynamics of a methylotroph *(Methylomonas L3)* in continuous cultures. I. Fast transients induced by methanol pulses and methanol accumulation, *Biotechnol. Bioeng.*, 29, 55, 1987.

33. **O'Neil, D. G. and Lyberatos, G.**, Feedback identification of continuous microbial growth systems, *Biotechnol. Bioeng.*, 28, 1323, 1986.

34. **Parkin, G. F. and Speece, R. E.**, Attached versus suspended growth anaerobic reactors: response to toxic substances, *Water Sci. Technol.*, 15, 261, 1983.

35. **Pickett, A. M.**, Growth in a changing environment, in *Microbial Population Dynamics*, Bazin, M. J., Ed., CRC Press, Boca Raton, FL, 1982, chap. 4.

Chapter 12

BIOLOGICAL TREATMENT: AN ENTREPRENEURIAL OPPORTUNITY

Franklin P. Johnson, Jr.

The death throes of birds and fish in the Kesterson Wildlife Refuge and the Carson Sink are the birth pangs of a new industry, the treatment of agricultural wastewater by biological means. The birth, however, will not be quick and it will not be easy.

The free enterprise system has provided fertile ground for the creation of many industries when entrepreneurs have coupled new science and technology to human needs. The process has been much the same from the days of the rough-hewn characters of the Monongahela Valley creating the steel industry 100 years ago to the externally smoother men and women of today's Silicon Valley. Whatever their time, the entrepreneurs have declared economic independence, scrambled for capital, made technology work, and battled for room in a marketplace they were helping to create. Beyond intellect and vision, they have had the essential qualities of zeal and courage, desire to succeed, and unwillingness to fail.

As entrepreneurs and innovative existing companies eye the market for treating agricultural wastewater, they run into a problem which is neither new nor common: there won't be any significant market until the government acts to create it.

It is clear that pesticides and herbicides, minerals leached from the soil, and fertilizers are leaving the farms via wastewater drains and entering the surface environment and the groundwater. It is also clear that plants and animals, including *Homo sapiens,* are (or will be) adversely affected. Significant percentages of California land will become useless because of salts concentrated through the groundwater cycle. There can be a large market with compelling economics, assuming modern bioscience and chemistry can economically capture these residues or degrade them into harmlessness or even usefulness.

The entrepreneurs must ask, however, exactly who will buy such products if resources and energy are invested into their development, manufacture, and marketing.

Ranchers and farmers will not be customers very soon because, in the absence of their acting in concert with all other farmers in their drainage basin, there is little return for their investment in treatment equipment and supplies. If society decides, acting through legislatures and the Congress, that water in farm drains must have certain qualities, farmers become a market on a schedule dictated by the legislation. Few can believe that such legislation could or should be passed without an exhaustive examination and the development of national policy on agriculture and the environment. There will also be an extended and noisy political battle about who should pay for what and the competitiveness of American agriculture.

Another future potential customer class will be districts, such as already existing irrigation districts or newly created wastewater disposal districts. Such districts, which have been governmental entities in the past, may in the future be privately owned corporations in the business of gathering and treating agricultural wastewater. These companies would be not only customers for suppliers of treatment equipment and supplies, but also an entrepreneurial opportunity in themselves and part of the emerging industry. The fact that they would use private investment, as opposed to government funding, would make the fiscal part of the political problems more tractable.

Whether the customer is a farmer, a group of farmers, or a treatment district or corporation, substantial markets won't exist until the government acts on the problem, and without substantial markets there will be no industry developed in the treatment of agricultural wastewater.

An industry does exist, of course, in the treatment of sewage and industrial effluents. There is also a potentially strong industry in the treatment of hazardous waste, both at existing dumps and in handling newly created waste. In addition, many companies exist which are using bioreactors utilizing natural bacteria, algae, or yeasts, as well as genetically altered organisms. These companies not only produce pharmaceuticals and vitamins, but also provide organisms for treating hazardous waste and sewage. Many of these companies will be alert to opportunities in agricultural wastewater, and several of them are young and energetic. When the agricultural wastewater market develops, it will be served in many cases by those companies already active in the hazardous waste and sewage treatment businesses.

Once legislation creates this nascent market, a great many entrepreneurial opportunities will exist for scientists and businessmen who understand the technical problems and the market. In order to be timely, some of the entrepreneurial action and risk taking will have to take place before there is any legislation in place. Some examples of opportunities may be:

1. Creation of herbicides, pesticides, and fertilizers in conjunction with specific downstream biotreatment for them, and their manufacture and distribution as total biochemical systems
2. Design, manufacture, and sale of equipment which samples incoming water and water under treatment and continually varies the treatment
3. Creation of service companies to gather and treat wastewater or groundwater in agricultural areas
4. Development and licensing of bioreactive systems to companies making and distributing treatment equipment

The best ideas will come from people who understand in depth the potential of the available technology and the needs of the marketplace. The best companies will come from people who can take these ideas and add the business disciplines of marketing, engineering, manufacturing, finance, control, and more marketing.

The nation clearly needs a solution to its problems of wastewater in a modern agricultural system, and the Congress must act. A legislative battle will rage and, in its resolution, major openings will occur.

Venture capital is available in large amounts, and experienced business people are looking for opportunities. The time in the sun for the entrepreneurial wastewater bioscientist is here — almost. It is in these hours of early dawn, however, that farsighted companies will be formed to prepare for the battles to come.

INDEX

A

Absorption of xenobiotics, 100
Acinetobacter, 57, 77, 84
AD, see Anaerobic digester
Adaptation, irreversible, 159
Adsorption, 19—20
  in algal systems, 98—100
  iron filing process, 42—43
Aeration, in algal raceway systems, 115—116
Aerobic bacteria
  nitroreductase in, 77
  selenium reduction by, 41
Aerobic fermentation, 48, 50
Agar, 118, 121
Agarose, 118, 121
Agricultural chemicals, degradation of, 82
Agricultural return flow, 16
Agricultural wastewater, see also Drainage water
  detoxification of, 84—85
  disposal districts, 164
  dissolved organic matter in, 96—97
  pollutants in, 16, 74
  trace metals in, 99—100
  treatment of, 16, 112, 115
    biological systems, 16—19
    chemical systems, 19—21
    decision-making process, 21—30
    entrepreneurial opportunities in, 163—164
    physical systems, 19—21
    problems, 149—152
      experimental approaches for reducing
        uncertainty, 159—160
      overall system behavior, 157—159
      systems view of uncertainties, 152—157
    treatment systems, algal
      dissolved organic matter, 96—97
      inorganic carbon intake, regulation of, 92—93
      nitrogen uptake pathways, 93—95
      N:P atomic ratio, importance of, 96
      phosphorus uptake pathways, 96
      trace metals, 97—100
      xenobiotics, removal of, 100—104

Air-lift reactors, 85
Alanine, 95
Alanine dehydrogenase, 95
Alaria marginata, 99
Alcaligenes, 57, 79
Alcaligenes eutrophus, 78
Alcohol fermentation, 48
Algae, see also Algal systems; Microalgae
  benefits of, 139—140
  bioflocculation of, 40—41
  by-products, 118, 121—122
  controlling growth of, 113, 115, 124
  as dependable reductant, 138
  disadvantages of, 116

eukaryotic, 94
growth rates, 112, 115
harvesting, 138
heterotrophy in, 118
hyperconcentrated cultures of, 117—118
inoculating cultures of, 113
mass culture of, 112
number of species, 112
precipitate removal from, 138
prokaryotic, 94
as selenium reductant, 133—136, 144
  costs, 136—138, 143—147
  production yields, 136—137
settling of, 112
"starter" cultures, 122—124
Algal ponds, see High-rate ponds
Algal systems, 18, 48, 112, see also Algae;
      Biological treatment systems; Microbial
      treatment systems
  benefits, 118, 124, 139—140
  by-products, 118, 121—122, 124
  closed culture, 116, 124
    advantages, 118
    costs, 118—119
    immobilized, 118—119
    mobilized, 117—118
  collections, 122—124
  costs, 116, 118—119, 143—147
  dissolved organic matter and, 96—97
  harvesting techniques, 119—121
  immobilized, 118—119
  mobilized, 117—118
  multispecies, 112—115
  open culture, 116, 124
    growth rate control, 113, 115
    multispecies, 112—115
    production yields, 112—113
    unialgal, 115
  optimum nutrient ratio, 96
  production yields, 112—113, 117—119
  raceways in, 115—116
  San Joaquin Valley, 40—41, 143—147
    algal-bacterial selenium removal process, 132—
      139
    algal systems, benefits of, 139—140
  selenium removal, 41, 143—147
    algal-bacterial process, 132—139
    algal systems, benefits of, 139—140
  "starter" cultures, 122—124
  unialgal, 115
  for wastewater treatment
    dissolved organic matter, 96—97
    inorganic carbon intake, regulation of, 92—93
    nitrogen uptake pathways, 93—95
    N:P atomic ratio, importance of, 96
    phosphorus uptake pathways, 96
    trace metals, 97—100
    xenobiotics, removal of, 100—104

target contaminants, 35—36
trace elements in, 5—6, 136
treatment approach, 34—35
treatment costs, 8, 143—147
treatments selected for, 38—43, 131—140
treatment technology options, 36—38
treatment of, 112, 115, 163—164
xenobiotics in, 144
Drains, artificial, 2
Dunaliella, 95
bioflocculation in, 120
as glycerol source, 139
in open cultures, 115
vitamin production in, 122
Dunaliella salina, vitamin production in, 122

E

Ecological confoundation factor, 155
Ecological level uncertainty, in microbial treatment
systems, 154—155, 158
Ecology, microbial, 68
EEB, see Environmental engineering biotechnology
Egregia menziesii, 98
Electrical conductivity, of drainage water, 136
Electrochemical treatment systems, 43
Electrodes, ion-specific, 97
Electrodialysis, 8, 20, see also Desalinization
Electroflocculation, 119
Electron-exchange resins, 43
Endrin, 5
Energy consumption
of distillation processes, 21
for high-rate ponds, 137
for membrane processes, 20
Engineering, see also Environmental engineering
genetic, 51, 54
uncertainty in, 149—152
experimental approaches for reducing, 159—160
overall system behavior and, 157—159
systems view of, 152—157
Entrepreneurial opportunities, in waste treatment
systems, 163—164
Environmental engineering, 67—68, 144
biotechnology in, 47—51, 54, 68—70, 144
closed algal system control and, 116, 118
Environmental engineering biotechnology (EEB),
47—51, 144
deterministic, 151—152, 158
predictive, 151—152
uncertainty in, 149—152
experimental approaches for reducing, 159—160
overall system behavior and, 157—159
systems view of, 152—157
Environmental factors
irrigation and, 2
in treatment selection, 24—25
Enzymes
carbon-fixing, 92—93
plasmalemma, 100

protein chemistry and, 156—157
Escherichia coli, 55—56
Ethanolic fermentation, 49
Ethylbenzene, 80
Ethylbromide, 77
s-Ethyl-*N,N*-dipropylthiocarbamate, 82
Ethylene dibromide, 77
Eucheuma spinosum, 98
Eucheuma striatum, 98
Eutrophication, 7, 34, 38, 48
Evaporation, 11, 21, 139—140
Extrachromosomal DNA, 157

F

Fermentation, see also Methane fermentation
aerobic, 48
alcohol, 48
anaerobic, 48
ethanolic, 49
Ferric sulfate, selenium reduction and, 41
Ferrodoxin, 93—95, 97
Fertilizer degradation, entrepreneurial opportunities
in, 163—164
Filtration, 19—20
algal harvesting by, 119—120, 140
filter feeders, 120
invertebrate, 121
vacuum, 119
Fixation
carbon, 92—93, 98
nitrogen, 93—95
Fixed-bed reactors, 48, 85
Fixed-film bacterial systems, 17—18
Fixed potential amperometry, 97
"Flashing light effect", 115
Flavin mononucleotide, 100
Flavobacterium
soil detoxification by, 83
water detoxification by, 84—85
Flocculation, 19—20, 118—120
Flotation, algal harvesting by, 119—120
Fluidized-bed reactors, 48, 85
Fluorene, 80
Food chain treatment process, 18—19, 121
Fortuitous metabolism, 77
Fucoidin, 98
Fucus vesiculosus, 99
Fulvic acids, 98
Fungicides, 74
Fungi, volatilization of selenium by, 41
Furans, halogenated, 51

G

Genes, 156—157
cloning, 48, 51
probing, 57—64, 159
Genetic engineering, 51, 54, 55—57
Genetics, 157

absorption, 100
algal removal of, 100—104
bioconcentration of, 100, 103—104
catabolism of, 157
decontamination of, 84—85
degradation, 74—82, 100, 102
desorption of, 100

Xylene, 79—80
  anaerobic bacterial processing of, 50—51
  degradation of, 57

Z

Zinc, 42, 97, 100

Milton Keynes UK
Ingram Content Group UK Ltd.
UKHW052017071024
449327UK00027B/2306